Introduction to Engineering Design & Problem Solving

McGraw-Hill's BEST—Basic Engineering Series and Tools

G. Bertoline, *Introduction to Graphics Communications for Engineers*

D. M. Burghardt, *Introduction to Engineering Design and Problem Solving*

S. Chapman, *Introduction to Fortran 90/95*

K. Donaldson, *The Engineering Student Survival Guide*

A. Eide, et al., *Introduction to Engineering Design and Problem Solving*

A. Eisenberg, *A Beginner's Guide to Technical Communication*

B. Gottfried, *Spreadsheet Tools for Engineers:* Excel 97 Version

B. Gottfried, *Spreadsheet Tools for Engineers:* Excel 2000 Version

R. Greenlaw and E. Hepp, *Introduction to the Internet for Engineers*

W. Palm, *Introduction to Matlab 6 for Engineers*

R. Pritchard, *Mathcad: A Tool for Engineering Problem Solving*

R. Schinzinger and M. W. Martin, *Introduction to Engineering Ethics*

K. Smith, *Project Management and Teamwork*

A. Tan and T. D'Orazio, *C Programming for Engineering and Computer Science*

Introduction to Engineering Design & Problem Solving

Arvid R. Eide
Iowa State University

Roland D. Jenison
Iowa State University

Lane H. Mashaw
Iowa State University

Larry L. Northup
Iowa State University

Second Edition

Boston   Burr Ridge, IL   Dubuque, IA   Madison, WI   New York   San Francisco   St. Louis
Bangkok   Bogotá   Caracas   Kuala Lumpur   Lisbon   London   Madrid   Mexico City
Milan   Montreal   New Delhi   Santiago   Seoul   Singapore   Sydney   Taipei   Toronto

McGraw-Hill Higher Education ⚛

*A Division of The **McGraw-Hill** Companies*

INTRODUCTION TO ENGINEERING DESIGN AND PROBLEM SOLVING
SECOND EDITION

Published by McGraw-Hill, a business unit of The McGraw-Hill Companies, Inc., 1221 Avenue of the Americas, New York, NY 10020. Copyright © 2002, 1998 by The McGraw-Hill Companies, Inc. All rights reserved. No part of this publication may be reproduced or distributed in any form or by any means, or stored in a database or retrieval system, without the prior written consent of The McGraw-Hill Companies, Inc., including, but not limited to, in any network or other electronic storage or transmission, or broadcast for distance learning.

Some ancillaries, including electronic and print components, may not be available to customers outside the United States.

This book is printed on acid-free paper.

2 3 4 5 6 7 8 9 0 QPF/QPF 0 9 8 7 6 5 4 3 2

ISBN 0–07–240221–0

General manager: *Thomas E. Casson*
Publisher: *Elizabeth A. Jones*
Executive editor: *Eric M. Munson*
Developmental editor: *Maja Lorkovic*
Marketing manager: *Ann Caven*
Project manager: *Jill R. Peter*
Production supervisor: *Enboge Chong*
Coordinator of freelance design: *Rick D. Noel*
Senior photo research coordinator: *Lori Hancock*
Senior supplement producer: *Audrey A. Reiter*
Media technology senior producer: *Phillip Meek*
Compositor: ***TECH**BOOKS*
Typeface: *10/12 Palatino*
Printer: *Quebecor World Fairfield, PA*

Library of Congress Cataloging-in-Publication Data

Introduction to engineering design and problem solving / Arvid R. Eide . . . [et al.].—2nd ed.
    p. cm.
  Revised ed. of: Introduction to engineering design, © 1998, and Introduction to engineering problem solving, © 1998.
    Includes index.
    ISBN 0–07–240221–0
    1. Engineering mathematics. 2. Engineering design. I. Eide, Arvid R. II. Title:
Introduction to engineering design. III. Title: Introduction to engineering problem solving.

TA330 .I64   2002
620'.0042—dc21   2001026615
                                                                   CIP

www.mhhe.com

About the Authors

Arvid R. Eide received his baccalaureate degree in mechanical engineering from Iowa State University. Upon graduation he spent two years in the U.S. Army as a commissioned officer and then returned to Iowa State as an instructor while completing a master's degree in mechanical engineering. Professor Eide has worked for Western Electric, John Deere, and the Trane Company. He received his Ph.D. in 1974 and was appointed professor and Chair of Freshman Engineering, a position he held from 1974 to 1989, at which time Dr. Eide was appointed Associate Dean of Academic Affairs. From 1996 to 1999 he returned to teaching as a professor of mechanical engineering. In January 2000 he retired from Iowa State University as professor emeritus of mechanical engineering.

Roland D. (Rollie) Jenison has 35 years of teaching experience in aerospace engineering and lower division general engineering. He has taught engineering problem solving, engineering design graphics, aircraft performance, and aircraft stability and control courses, in addition to serving as academic adviser to many engineering students. He is a longtime member of the American Society for Engineering Education (ASEE) and has published many papers on engineering education. He served as Chair of the Engineering Design Graphics Division of ASEE in 1986–1987. He has been active in the development of improved teaching methodologies through the application of team learning, hands-on projects, and open-ended problem solving. He retired in June 2000 as professor emeritus in the Department of Aerospace Engineering and Engineering Mechanics at Iowa State University.

Lane H. Mashaw earned his BSCE from the University of Illinois and MSCE from the University of Iowa. He served as a municipal engineer in Champaign, IL, Rockford, IL, and Iowa City, IA, for nine years and then was in private practice in Decatur, IL, for another nine years. He taught at the University of Iowa from 1964 to 1974 and at Iowa State University from 1974 until his retirement in 1987. He is currently professor emeritus of civil and construction engineering.

Larry L. Northup is a professor of civil and construction engineering at Iowa State University. He has more than 35 years of teaching experience, with the past 25 years devoted to lower division engineering courses in problem solving, graphics, and design. He has 2 years of industrial experience and is a registered engineer in Iowa. He has been active in ASEE (Engineering Design Graphics Division), having served as Chair of the Freshman Year Committee and Director of Technical and Professional Committees (1981–1984). He also served as Chair of the Freshman Programs Constituent Committee (now Division) of ASEE in 1983–1984.

Contents

Preface   viii

1   The Engineering Profession   1

1.1   Introduction   1
1.2   The Technology Team   2
1.3   The Functions of the Engineer   11
1.4   The Engineering Disciplines   23
1.5   Total Quality in Engineering   34
1.6   Background and History   36
1.7   Definitions and Tools   41
1.8   Problem Solving and the Team   47
1.9   Education of the Engineer   61
1.10   Professionalism and Ethics   66
1.11   Challenges of the Future   69

2   Engineering Design—A Process   79

2.1   Introduction   79
2.2   Identification of a Need—Step 1   86
2.3   Problem Definition—Step 2   88
2.4   Search—Step 3   90
2.5   Constraints—Step 4   94
2.6   Criteria—Step 5   95
2.7   Alternative Solutions—Step 6   98
2.8   Analysis—Step 7   102
2.9   Decision—Step 8   108
2.10   Specification—Step 9   113
2.11   Communication—Step 10   115

3   Engineering Solutions   125

3.1   Introduction   125
3.2   Problem Analysis   125
3.3   The Engineering Method   126
3.4   Problem Presentation   128
3.5   Standards of Problem Presentation   129

4 Representation of Technical Information   145

4.1   Introduction   145
4.2   Collecting and Recording Data   149
4.3   General Graphing Procedures   151
4.4   Empirical Functions   163
4.5   Curve Fitting   163
4.6   Method of Selected Points and Least Squares   164
4.7   Empirical Equations—Linear   164
4.8   Empirical Equations—Power Curves   166
4.9   Empirical Equations—Exponential Curves   172

5 Engineering Estimations and Approximations   183

5.1   Introduction   183
5.2   Significant Digits   183
5.3   Accuracy and Precision   187
5.4   Errors   188
5.5   Approximations   190

6 Dimensions, Units, and Conversions   201

6.1   Introduction   201
6.2   Physical Quantities   201
6.3   Dimensions   202
6.4   Units   204
6.5   SI Units and Symbols   205
6.6   Rules for Using SI Units   210
6.7   U.S. Customary and Engineering System   215
6.8   Conversion of Units   217

Answers to Selected Problems   227

Index   229

Preface

To the Student

As you begin the study of engineering no doubt you are filled with enthusiasm, curiosity, and a desire to succeed. Your first year will be spent primarily establishing a solid foundation in mathematics, basic sciences, and communications. At times you may question what the benefits of this background material are and when actual engineering experiences will begin. We believe that those engineering experiences begin now. As authors, we believe that the material presented in this book will provide you with a fundamental understanding of how engineers function in today's technological world. To accomplish this, we apply engineering principles to the solution of problems.

To the Instructor

Engineering courses for first-year students continue to be in a state of transition. A broad set of course goals, including coverage of prerequisite material, motivation and retention have spawned a variety of first-year activity. The traditional engineering drawing and descriptive geometry courses have been largely replaced with computer graphics and CAD-based courses. Courses in introductory engineering and problem solving are now utilizing spreadsheets and mathematical solvers in addition to teaching the rudiments of a computer language. The World Wide Web (WWW) has become a major instructional tool, providing a wealth of data to supplement your class notes and textbooks. This edition continues the authors' intent to introduce the profession of engineering and provide students with many of the tools needed to succeed.

Since 1974 students at Iowa State University have taken a computations course that has a major objective of improving their problem-solving skills. Various computational aids have been used from programmable calculators to network PCs. This BEST edition of the text has evolved from nearly 30 years of experience with teaching engineering problem solving to thousands of first-year students.

This edition has the same broad objectives as the previous editions: (1) to motivate engineering students during their first year when exposure to the subject matter of engineering is limited; (2) to provide students with experience in solving problems in both SI and customary units while presenting solutions in a logical manner; (3) to develop students' skills in solving open-ended problems and (4) to provide fundamental problem solving skills that will serve students their entire engineering profession.

The material in this book is presented in a manner that allows each instructor to emphasize certain aspects more than others without loss of continuity. The problems that follow most chapters vary in difficulty, so that students can experience success rather quickly and still be challenged as problems become more complex. Most problems in this edition are new; however, some have been modified and there are more computer-oriented problems.

The book has been conveniently divided into two major sections. The first, an introduction to engineering, begins with a description and breakdown of the engineering profession. Material concerning most disciplines in engineering is included in this edition. If a formal orientation course is given separately, Chapter 1 can be simply a reading assignment and the basis for students to investigate disciplines of interest. Chapter 1 includes a brief introduction to the principles of total quality management (TQM). Students are introduced to the importance of being competitive in the international marketplace and how to use teaming and problem-solving skills to achieve a high level of competitiveness. Engineering design is introduced in Chapter 2, providing an opportunity to investigate the "essence of engineering" in a holistic manner. By studying the chapter example illustrating the steps in a structured design process, students will understand the need for developing their technical and nontechnical capabilities in their chosen discipline. Design is treated as a cognitive activity involving the processing of vast amounts of information, critical thinking, and decision making in a team environment. Students are best served in an introduction to design by having them work on a team to produce a conceptual design to satisfy a stated need.

The second major section, processing engineering data, includes materials we believe that all engineering students require in preparation for any engineering curriculum. Chapters 3 and 4 provide procedures for approaching an engineering problem, determining the necessary data and method of solution, and presenting the results.

Chapters 5 and 6 in this edition include engineering estimations, and dimensions and units (including both customary and SI units). Throughout the book, discussions and example problems tend to emphasize SI metric. However, other dimension systems still remain in use today, so a number of our examples and problems contain nonmetric units to ensure that students are exposed to conversions and other units that are commonly used.

Since the text was written for first-year engineering students, mathematical expertise beyond algebra, trigonometry and analytical geometry is not required for any material in the book. The authors have found, however, that additional experience in pre-calculus mathematics is very helpful as a prerequisite for this text.

A solutions manual is available that contains solutions to most end-of-chapter problems. Please contact your local or regional McGraw-Hill representative.

Acknowledgments

The authors are indebted to many who assisted in the development of this edition of the textbook. First, we would like to thank the faculty of the former Division of Engineering Fundamentals and Multidisciplinary Design at Iowa State University who have taught the engineering computations courses over the past 25 years. They, with the support of engineering faculty from other departments, have made the courses a success by their efforts. Several thousands of students have taken the courses, and we want to thank them for their comments and ideas that have influenced this edition. The many suggestions of faculty and students alike have provided us with much information that was necessary to improve the previous editions. A special thanks goes to the reviewers for this edition whose suggestions were extremely valuable. These suggestions greatly shaped the manuscript in preparation of this edition.

The authors would like to thank particularly Rebecca Sidler Kellogg for the material in Chapter 2 on modern design principles and research into the cognitive nature of engineering design. Efforts to develop critical thinking skills, project management capability, and other skills necessary for success in engineering are supported greatly by involvement in design. The ability to study design as a logical process from the very beginning of an engineering curriculum is a major accomplishment in engineering education.

The authors also thank the following graduate and undergraduate students for problem suggestions and solutions: Scott Openshaw, Samyukta Sankaran, Jeff Vanek, Sarah Allen, Brian Jensen, and Jesse Hilton.

Finally, we thank our families for their constant support of our efforts.

Arvid R. Eide
Roland D. Jenison
Lane H. Mashaw
Larry L. Northup

CHAPTER 1

The Engineering Profession

1.1   Introduction

The rapidly expanding sphere of science and technology may seem overwhelming to the individual seeking a career in a technological field. A technical specialist today may be called either engineer, scientist, technologist, or technician, depending on education, industrial affiliation, or specific work. For example, nearly 350 colleges and universities offer engineering programs accredited by the Accreditation Board for Engineering and Technology (ABET) or the Canadian Engineering Accreditation Board (CEAB). Included in these programs are such traditional specialities as aerospace, agricultural, chemical, civil, computer, construction, electrical, industrial, and mechanical engineering — as well as the expanding areas of energy, environmental, and materials engineering. Programs in engineering science, mining engineering, and petroleum engineering add to a lengthy list of career options in engineering alone. Coupled with thousands of programs in science and technical training offered at hundreds of universities, colleges, and technical schools, the task of choosing the right field no doubt seems formidable.

Since you are reading this book, we assume that you are interested in studying engineering or at least are trying to decide whether or not to do so. Up to this point in your academic life you probably have had little experience with engineering and have gathered your impressions of engineering from advertising materials, counselors, educators, and perhaps a practicing engineer or two. Now you must investigate as many careers as you can as soon as possible to be sure of making the right choice.

The study of engineering requires a strong background in mathematics and the physical sciences. Section 1.9 discusses typical areas of study within an engineering program that lead to the bachelor's degree. You also should consult with your academic counselor about specific course requirements. If you are enrolled in an engineering program but have not chosen a specific discipline, consult with an adviser or someone on the engineering faculty about particular course requirements in your areas of interest.

When considering a career in engineering or any closely related fields, you should explore the answers to several questions. What is engineering? What is an engineer? What are the functions of engineering? What are the engineering disciplines? Where does the engineer fit into the technical spectrum? How are engineers educated? What is meant by professionalism and

Figure 1.1

Imagine the number of engineers who were involved in the design and construction of the tanker, the pumping system for unloading, the storage facilities, and the port facilities for this imported oil unloading station.

engineering ethics? What have engineers done in the past? What are engineers doing now? What will engineers do in the future? Finding answers to such questions will assist you in assessing your educational goals and in obtaining a clearer picture of the technological sphere.

Brief answers to some of these questions are given in this chapter. By no means are they intended to be a complete discussion of engineering and related fields. You can find additional and more detailed technical career information in the reference materials listed in the bibliography at the end of the book and by searching the World Wide Web (WWW).

1.2 The Technology Team

In 1876, 15 men led by Thomas Alva Edison gathered in Menlo Park, New Jersey, to work on "inventions." By 1887, the group had secured over 400 patents, including ones for the electric light bulb and the phonograph. Edison's approach typified that used for early engineering developments. Usually one person possessed nearly all the knowledge in one field and directed the research, development, design, and manufacture of new products in this field.

Today, however, technology has become so advanced and sophisticated that one person cannot possibly be aware of all the intricacies of a single device or process. The concept of systems engineering thus has evolved; that is, technological problems are studied and solved by a technology team.

Scientists, engineers, technologists, technicians, and craftspersons form the *technology team.* The functions of the team range across what often is called the *technical spectrum.* At one end of the spectrum are functions which involve work with scientific and engineering principles. At the other end of this technical spectrum are functions which bring designs into reality. Successful technology teams use the unique abilities of all team members to bring about a successful solution to a human need.

Each of the technology team members has a specific function in the technical spectrum, and it is of utmost importance that each specialist understand the role of all team members. It is not difficult to find instances where the education and tasks of team members overlap. For any engineering accomplishment successful team performance requires cooperation that can be realized only through an understanding of the functions of the technology team. The technology team is one part of a larger team which has the overall responsibility for bringing a device, process, or system into reality. This team, frequently called a project or design team, may include, in addition to the technology team members, managers, sales representatives, field service persons, financial representatives, and purchasing personnel. These project teams meet frequently from the beginning of the project to ensure that schedules and design specifications are met, and that potential problems are diagnosed early. This approach, intended to meet or exceed the customer's expectations, is referred to as total quality management (TQM) or continuous improvement (CI). Total quality is introduced in Sec. 1.4. We will now investigate each of the team specialists in more detail.

1.2.1 Scientist

Scientists have as their prime objective increased knowledge of nature (see Fig. 1.2). In the quest for new knowledge, the scientist conducts research in a systematic manner. The research steps, referred to as the *scientific method,* are often summarized as follows:

1. Formulate a hypothesis to explain a natural phenomenon.
2. Conceive and execute experiments to test the hypothesis.
3. Analyze test results and state conclusions.
4. Generalize the hypothesis into the form of a law or theory if experimental results are in harmony with the hypothesis.
5. Publish the new knowledge.

An open and inquisitive mind is an obvious characteristic of a scientist. Although the scientist's primary objective is that of obtaining an increased knowledge of nature, many scientists are also engaged in the development of their ideas into new and useful creations. But to differentiate quite simply between the scientist and engineer, we might say that the true scientist seeks to understand more about natural phenomena, whereas the engineer primarily engages in applying new knowledge.

Figure 1.2

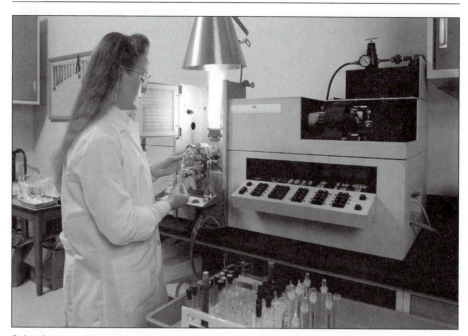

Scientists use the laboratory for discovery of new knowledge.

1.2.2 Engineer

The profession of engineering takes the knowledge of mathematics and natural sciences gained through study, experience, and practice and applies this knowledge with judgment to develop ways to utilize the materials and forces of nature for the benefit of all humans.

An engineer is a person who possesses this knowledge of mathematics and natural sciences, and through the principles of analysis and design, applies this knowledge to the solution of problems and the development of devices, processes, structures, and systems for the benefit of all humans.

Both the engineer and scientist are thoroughly educated in the mathematical and physical sciences, but the scientist primarily uses this knowledge to acquire new knowledge, whereas the engineer applies the knowledge to design and develops usable devices, structures, and processes. In other words, the scientist seeks to know, the engineer aims to do.

You might conclude that the engineer is totally dependent on the scientist for the knowledge to develop ideas for human benefit. Such is not always the case. Scientists learn a great deal from the work of engineers. For example, the science of thermodynamics was developed by a physicist from studies of practical steam engines built by engineers who had no science to guide them. On the other hand, engineers have applied the principles of nuclear fission discovered by scientists to develop nuclear power plants and numerous other devices and systems requiring nuclear reactions for their operation. The scientist's and engineer's functions frequently overlap, leading at times to a somewhat blurred image of the engineer. What distinguishes the engineer from the scientist in broad terms, however, is that the engineer often conducts research but does so for the purpose of solving a problem.

The end result of an engineering effort—generally referred to as *design*—is a device, structure, system, or process which satisfies a need. A successful design is achieved when a logical procedure is followed to meet a specific need. The procedure, called the *design process,* is similar to the scientific method with respect to a step-by-step routine, but it differs in objectives and end results. The design process encompasses the following activities, all of which must be completed.

1. Identification of a need
2. Problem definition
3. Search
4. Constraints
5. Criteria
6. Alternative solutions
7. Analysis
8. Decision
9. Specification
10. Communication

In the majority of cases designs are not accomplished by an engineer who is simply completing the 10 steps shown in the order given. As the designer proceeds through each step, new information may be discovered and new objectives may be specified for the design. If so, the designer must backtrack and repeat steps. For example, if none of the alternatives appears to be economically feasible when the final solution is to be selected, the designer must redefine the problem or possibly relax some of the criteria to admit less expensive alternatives. Thus, because decisions must frequently be made at each step as a result of new developments or unexpected outcomes, the design process becomes iterative.

It is very important that you begin your engineering studies with an appreciation of the thinking process used to arrive at a solution to a problem and ultimately to produce a successful result.

As you progress through your engineering education, you will solve problems and learn the design process using the techniques of analysis and synthesis. Analysis is the act of separating a system into its constituent parts, whereas synthesis is the act of combining parts into a useful system. In the design process (Chapter 2) you will observe how analysis and synthesis are utilized to generate a solution to a human need.

Consider the cruise control in an automobile as a system. You can analyze the performance of this system by setting up a test under carefully controlled conditions; that is, you will define and control the operating environment for the system and note the performance of the system. For example, you may determine acceleration or deceleration when a speed change is requested by the driver. You may check to see if the speed returns to the desired level after braking to reduce the speed and using the resume control. During the design of a cruise control system the system would be modeled on a computer and performance would be predicted by adjusting the variables and observing the results through various graphical formats on the monitor. You can analyze the physical makeup of the cruise control by actually taking apart the control, identifying the parts according to form and function, and reassembling the control. In general, analysis is taking a system, establishing the operating environment, and determining the response (performance) of the system.

If you were attempting to design a new cruise control system, you would consider many methods for sensing speed, ways to adjust engine speed for acceleration and deceleration, ideas for driver interface with the control, and so forth. Many possible solutions will be generated, mostly in the form of conceptual solutions without the details. During the design phase the computer model may be continually improved by "repeated analysis," that is, finding the best or optimum design by observing the effect of changes in the system variables. This is synthesis, the inverse of analysis. Synthesis may be stated as the process of defining the desired response (performance) of a system, establishing the operating environment, and, from this, developing the system.

Three examples will further illustrate analysis and synthesis.

Figure 1.3

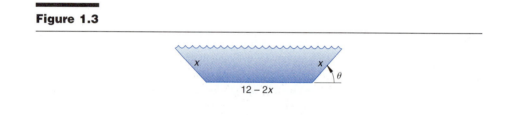

$$12 - 2x$$

Example problem 1.1   A protective liner exactly 12 m wide is available to line a channel for conveying water from a reservoir to downstream areas. If a trapezoidal-shaped channel (see Fig. 1.3) is constructed so that the liner will cover the surface completely, what is the flow area for $x = 2$ m and $\theta = 45°$? The geometry is defined such that $0 \leq x \leq 6$ and $0 \leq \theta \leq 90°$. Flow area multiplied by average flow velocity will yield volume rate of flow, an important parameter in the study of open-channel flows.

Solution   The geometry is defined in Fig. 1.3. The flow area is given by the expression for the area of a trapezoid.

$$A = \frac{1}{2}(b_1 + b_2)h$$

where $b_1 = 12 - 2x$

$b_2 = 12 - 2x + 2x \cos \theta$

$h = x \sin \theta$

Therefore

$$A = 12x \sin \theta - 2x^2 \sin \theta + x^2 \sin \theta \cos \theta$$

For the situation where $x = 2$ and $\theta = 45°$, the flow area is

$$A = (12)(2)(\sin 45°) - 2(2)^2(\sin 45°) + (2)^2(\sin 45°)(\cos 45°)$$

$$= 13.3 \text{ m}^2$$

Values of A can be quickly found for any combination of x and θ with a spreadsheet. Figure 1.4a shows areas for $x = 2$ and a series of θ values. You have solved many problems of this nature by analysis; that is, a system is given (the channel as shown in Fig. 1.3), the operating environment is specified (the channel is flowing full), and you must find the system performance (determine the flow area). Analysis usually yields a unique solution.

Example problem 1.2 A protective liner exactly 12 m (meters) wide is available to line a channel conveying water from a reservoir to downstream areas. For the trapezoidal cross section shown in Fig. 1.3, what are the values of x and θ for a flow area of 16 m^2?

Solution Based on our work in Example prob. 1.1, we have

$$16 = 12x \sin \theta - 2x^2 \sin \theta + x^2 \sin \theta \cos \theta$$

The solution procedure is not direct, and the solution is not unique, as it was in Example prob. 1.2. We begin our solution procedure by using a spreadsheet to generate a family of curves that illustrate the behavior of the implicit function of x and θ. Figure 1.4b shows the flow area as a function of θ for five values of x. We quickly observe that for $x = 1$ we cannot generate a flow area of 16 m^2. Also, for $x = 4$ and 5 we definitely have two values of θ where a flow of 16 m^2 is possible. The spreadsheet will perform a search for the correct values. Figure 1.4c shows the result of a search between 0 and 90 at $x = 3$ m. In this situation a flow area of 16 m^2 occurs at a θ of 0.698014 radians or 40.0°. Other results are quickly obtainable by simply changing the value of x in the spreadsheet program in Fig. 1.4c.

You probably have not solved many problems of this nature. Example prob. 1.2 is a synthesis problem; that is, the operating environment is specified (channel flowing full), the performance is known (flow area is 16 m^2), and you must determine the system (values for x and θ). Example prob. 1.2 is the inverse problem to Example prob. 1.1. In general, synthesis problems do not have a unique solution, as can be seen from Example prob. 1.2.

Most of us have difficulty synthesizing. We cannot "see" a direct method to find an x and θ that yield a flow area of 16 m^2. Our solution to Example prob. 1.2 involved repeated analysis to "synthesize" the solution. We studied a family of curves (x is constant) of A versus θ, which enabled us to verify the spreadsheet analysis for values of x and θ that yield a specified flow area.

Example problem 1.3 For the situation described in Example prob. 1.2, find values of x and θ that yield a maximum flow area.

Figure 1.4

	x		Theta, D		Theta, R		Area
	2		0		0		0
	2		15		0.261799		5.141105
	2		30		0.523599		9.732051
	2		45		0.785398		13.31371
	2		60		1.047198		15.58846
	2		75		1.308997		16.45481
	2		90		1.570796		16
Area = $12*x*\sin(Theta)-2*x^2*\sin(Theta)+x^2*\sin(Theta)*\cos(Theta)$							

(*a*)

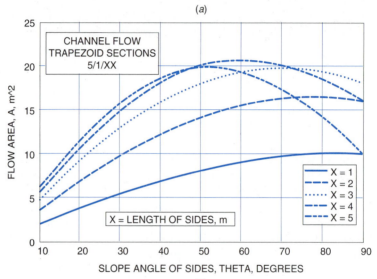

(*b*)

	x			Theta,D		Theta, R		Area
		3		0		0		0
				90		1.570 796		18
			Area = 16 For Theta =			0.698014 Radians		

(*c*)

Theta, Degree	Theta, Radians	x, m			
0	0	1			
15	0.261799	2			
35	0.610865	3			
55	0.959931	4			
75	1.308997	5			
90	1.570796	6			
Max flow area	Theta	x			Area
	1.047198	4			20.78461
Area = $12*x*\sin(Theta)-2*x^2*\sin(Theta)+x^2*\sin(Theta)*\cos(Theta)$					

(*d*)

Solution   This is a design problem in which a "best" solution is sought, in this case a maximum flow area for the trapezoidal cross section shown in Fig. 1.3. We can determine the solution from the repeated analysis we did for Example prob. 1.2. The solution is obtained readily from a spreadsheet search over the range of values for x and θ. Figure 1.4d shows the result. You can verify this by checking against Fig. 1.4b.

Analysis and synthesis are very important to the engineering design effort, and a majority of your engineering education will involve techniques of analysis and synthesis in problem solving. We must not, however, forget the engineer's role in the entire design process. In an industrial setting the objective is to assess correctly a need, to determine the best solution the need, and to market the solution more quickly and less expensively than the competition. This demands careful adherence to the design process.

The successful engineer in a technology team will take advantage of computers and computer graphics. Today, with the aid of computers and computer graphics, it is possible to perform analysis, to decide among alternatives, and to communicate results far more quickly and with more accuracy than ever before. This translates into better engineering and an improved quality of living.

Terms like *computer-aided design* (CAD) and *computer-aided manufacturing* (CAM) label the modern engineering activities that continue to make engineering a challenging profession. Your work with analysis and synthesis techniques will require the use of a computer to a large extent in your education.

In addition to the use of the computer to perform computations and to develop models, the information superhighway will provide you with instant access to new technologies, new processes, technical information, current economic conditions, and a multitude of other data that will help you to achieve success in the workplace and in all other aspects of life. The Internet, a worldwide collection of computer networks, now has over 3 million servers which can be accessed for information. The World Wide Web provides a user-friendly graphics interface to the Internet enabling text, audio, and video to be transmitted. Various search methods exist to help you find the information you are seeking.

Working engineers are now able to communicate with colleagues around the world via electronic mail (e-mail) on common interests and problems. We are able to monitor in real time a field test of our design in a foreign nation as we sit at our desks in the United States. With company electronic networks now available, called Intranets, databases of products, production status, design changes, and field status are at your fingertips.

In your personal life you are able to join user groups with common interests in sports, music, home maintenance, automobile repair, and the like. The new paradigm of online, interactive control of the information we desire is unprecedented: We can get what information we want and when we want it. Not since the Gutenberg press has such a dramatic change occurred in the way that we acquire and distribute knowledge.

Figure 1.5

Technicians check out complex electrical systems.

1.2.3   Technologist and Technician

Much of the actual work of converting the ideas of scientists and engineers into
tangible results is performed by technologists and technicians (see Fig. 1.5). A
technologist generally possesses a bachelor's degree and a technician an associate's degree. Technologists are involved in the direct application of their education and experience to make appropriate modifications in designs as the
need arises. Technicians primarily perform computations and experiments and
prepare design drawings as requested by engineers and scientists. Thus technicians (typically) are educated in mathematics and science but not to the depth
required of scientists and engineers. Technologists and technicians obtain a
basic knowledge of engineering and scientific principles in a specific field and
develop certain manual skills that enable them to communicate technically with
all members of the technology team. Some tasks commonly performed by technologists and technicians include drafting, estimating, model building, data
recording, and reduction, troubleshooting, servicing, and specification. Often
they are the vital link between the idea on paper and the idea in practice.

1.2.4   Skilled Trades/Craftspersons

Members of the skilled trades possess the skills necessary to produce parts
specified by scientists, engineers, technologists, and technicians. Craftspersons
do not need to have an in-depth knowledge of the principles of science and
engineering incorporated in a design (see Fig. 1.6). They often are trained on

Figure 1.6

Skilled craftspersons are the key elements in a manufacturing process.

the job, serving an apprenticeship during which the skills and abilities to build and operate specialized equipment are developed. Some of the specialized jobs of craftspersons include those of welder, machinist, electrician, carpenter, plumber, and mason.

1.3 The Functions of the Engineer

As we alluded to in the previous section, engineering feats accomplished from earliest recorded history up to the Industrial Revolution could best be described as individual accomplishments. The various pyramids of Egypt were usually designed by one individual, who directed tens of thousands of laborers during construction. The person in charge called every move, made every decision, and took the credit if the project was successful or the consequences if the project failed.

With the Industrial Revolution, there was a rapid increase in scientific findings and technological advances. One-person engineering teams were no longer practical or desirable. We know today that no single aerospace engineer is responsible for the jumbo jets and no one civil engineer completely designs a bridge. Automobile manufacturers assign several thousand engineers to the design of a new model. So we not only have the technology team as described earlier, but we have engineers from many disciplines who are working together on single projects.

One approach to an explanation of an engineer's role in the technology spectrum is to describe the different types of work that engineers do. For example, civil, electrical, mechanical, and other engineers become involved in design, which is an engineering function. The *engineering functions,* which are discussed briefly in this section, are research, development, design, production, testing, construction, operations, sales, management, consulting, and teaching. Several of the *engineering disciplines* will be discussed later in this section.

To avoid confusion between the meaning of the engineering disciplines and engineering functions, let us consider the following. Normally a student selects a curriculum (e.g., aerospace, chemical, mechanical, etc.) either before or soon after admission to an engineering program. When and how the choice is made varies with each school. The point is, that the student does not choose a function, but, rather a discipline. To illustrate further, consider a student who has chosen mechanical engineering. This student will, during an undergraduate education, learn how mechanical engineers are involved in the engineering functions of research, development, design, and so on. Some program options allow a student to pursue an interest in a specific subdivision within the curriculum, such as energy conversion in a mechanical engineering program. Most other curricula have similar options.

Upon graduation, when you accept a job with a company, you will be assigned to a functional team performing in a specific area such as research, design, or sales. Within some companies, particularly smaller ones, you may become involved in more than one function—design *and* testing, for example. It is important to realize that regardless of your choice of discipline, you may become involved in one or more of the functions to be discussed in the following paragraphs.

1.3.1 Research

Successful research is one catalyst for starting the activities of a technology team or, in many cases, the activities of an entire industry. The research engineer seeks new findings, as does the scientist; but one must keep in mind that the research engineer also seeks a way to use the discovery.

Key qualities of a successful research engineer are perceptiveness, patience, and self-confidence. Most students interested in research will pursue the master's and doctor's degrees in order to develop their intellectual abilities and the necessary research skills. An alert and perceptive mind is needed to recognize nature's truths when they are encountered. When attempting to reproduce natural phenomena in the laboratory, cleverness and patience are prime attributes. Research often involves tests, failures, retests, and so on for

Figure 1.7

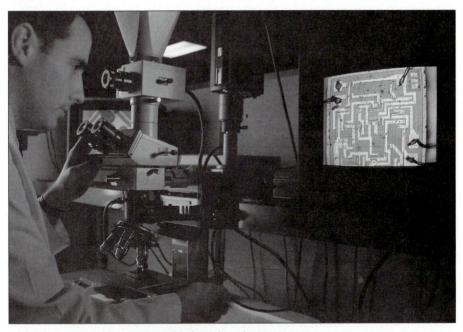

Research requires high-cost, sophisticated equipment.

long periods of time. Research engineers therefore are often discouraged and frustrated and must strain their abilities and rely on their self-confidence in order to sustain their efforts to a successful conclusion.

Billions of dollars are spent each year on research at colleges and universities, industrial research laboratories, government installations, and independent research institutes. The team approach to research is predominant today primarily because of the need to incorporate a vast amount of technical information into the research effort. Individual research also is carried out but not to the extent it was several years ago. A large share of research monies are channeled into the areas of energy, environment, health, defense, and space exploration. Research funding from federal agencies is very sensitive to national and international priorities. During a career as a research engineer you might expect to work in many diverse, seemingly unrelated areas, but your qualifications will allow you to adapt to many different research efforts.

1.3.2 Development

Using existing knowledge and new discoveries from research, the development engineer attempts to produce a device, structure, or process that is functional. Building and testing scale or pilot models is the primary means by which the development engineer evaluates ideas. A major portion of the development work requires use of well-known devices and processes in conjunction with established theories. Thus reading available literature and having a solid background

Figure 1.8

Development engineers take an idea and produce a concept of a functional product or system. The result of this activity is passed on to the design engineers for completing necessary details for production.

in the sciences and in engineering principles are necessary for the development of the engineer's success.

Many people who suffer from heart irregularities are able to function normally today because of the pacemaker, an electronic device that maintains a regular heartbeat. The pacemaker is an excellent example of the work of development engineers.

The first model, conceived by medical personnel and developed by engineers at the Electrodyne Company, was an external device that sent pulses of energy through electrodes to the heart. However, the power requirement for stimulus was so great that patients suffered severe burns on their chests. As improvements were being studied, research in surgery and electronics enabled development engineers to devise an external pacemaker with electrodes through the chest attached directly to the heart. Although more efficient from the standpoint of power requirements, the devices were uncomfortable, and patients frequently suffered infection where the wires entered the chest. Finally, two independent teams developed the first internal pacemaker, 8 years after the original pacemaker had been tested. Their experience and research with tiny pulse generators for spacecraft led to this achievement. But the very fine wire used in these early models proved to be inadequate and quite often failed, forcing patients to have the entire pacemaker replaced. A team of engineers at General Electric developed a pacemaker that incorporated a new wire, called a *helicable*. The helicable consisted of 49 strands of wire coiled together and then wound into a spring. The spring diameter was about 46 μm (micrometers), one-half the diameter of a human hair. Thus, with doctors and development engineers working together, an effective, comfortable device was perfected that has enabled many heart patients to enjoy a more active life. Today pacemakers have been developed that operate at more than one speed, enabling the patient to speed up or slow down heart rate depending on physical activity. In addition to the advances in pacemakers, medical research has evolved many other procedures for correcting heart deficiencies. The field of electrophysiology, combining cardiology with electrical and computer engineering, is enabling thousands of persons with heart irregularities to live productive and happy lives.

We have discussed the pacemaker in detail to point out that changes in technology can be in part owing to development engineers. Only 13 years to develop an efficient, dependable pacemaker; 5 years to develop the transistor; and 25 years to develop the digital computer indicate that modern engineering methods generate and improve products nearly as fast as research generates new knowledge.

Successful development engineers are ingenious and creative. Astute judgment often is required in devising models that can be used to determine whether a project will be successful in performance and economical in production. Obtaining an advanced degree is helpful, but not as important as it is for an engineer who will be working in research. Practical experience more than anything else produces the qualities necessary for a career as a development engineer.

Development engineers frequently are asked to demonstrate that an idea will work. Within certain limits they do not work out the exact specifications that a final product should possess. Such matters are usually left to the design engineer if the idea is deemed feasible.

1.3.3 Design

The development engineer produces a concept or model that is passed on to the design engineer for converting into a device, process, or structure (see Fig. 1.9). The designer relies on education and experience to evaluate many possible solutions, keeping in mind the cost of manufacture, ease of production, availability of materials, and performance requirements. Usually several designs and redesigns will be undertaken before the product is brought before the general public.

To illustrate the role that the design engineer plays, we will discuss the development of the over-the-shoulder seat belts for added safety in automobiles, which created something of a design problem. Designers had to decide where and how the anchors for the belts would be fastened to the car body. They had to determine what standard parts could be used and what parts had to be designed from scratch. Consideration was given to passenger comfort, inasmuch as awkward positioning could deter usage. Materials to be used for the anchors and the belt had to be selected. A retraction device had to be designed that would give flawless performance.

Figure 1.9

A team of design engineers review a proposed design solution.

From one such part of a car, an individual can extrapolate the numerous considerations that must be given to approximately 12 000 other parts that form the modern automobile: Optimum placement of engine accessories, comfortable design of seats, maximization of trunk space, and aesthetically pleasing body design all require thousands of engineering hours to be successful in a highly competitive industry.

Like the development engineer, the designer is creative. However, unlike the development engineer, who is usually concerned only with a prototype or model, the designer is restricted by the state of the art in engineering materials, production facilities, and, perhaps most important, economic considerations. An excellent design from the standpoint of performance may be completely impractical when viewed from a monetary point of view. To make the necessary decisions, the designer must have a fundamental knowledge of many engineering specialty subjects as well as an understanding of economics and people.

Figure 1.10

17

*The Functions of
the Engineer*

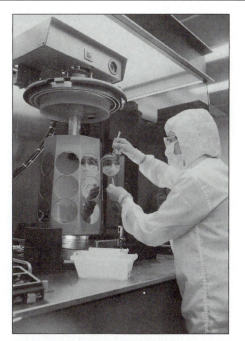

Test engineering is a major component in the development of new products.

1.3.4 Production and Testing

When research, development, and design have created a device for use by the public, the production and testing facilities are geared for mass production (see Figs. 1.10 and 1.11). The first step in production is to devise a schedule that will efficiently coordinate materials and personnel. The production engineer is responsible for such tasks as ordering raw materials at the optimum times, setting up the assembly line, and handling and shipping the finished product. The individual who chooses this field must possess the ability to visualize the overall operation of a particular project as well as know each step of the production effort. Knowledge of design, economics, and psychology is of particular importance for production engineers.

Test engineers work with a product from the time it is conceived by the development engineer until such time as it may no longer be manufactured. In the automobile industry, for example, test engineers evaluate new devices and materials that may not appear in automobiles until several years from now. At the same time they test component parts and completed cars currently coming off the assembly line. They are usually responsible for quality control of the manufacturing process. In addition to the education requirements of the design and production engineers, a fundamental knowledge of statistics is beneficial to the test engineer.

Figure 1.11

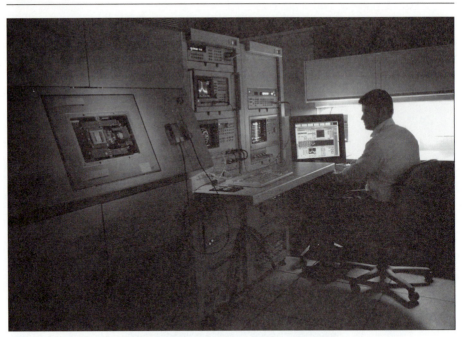

Modern production facilities are often controlled from a central computer system.

1.3.5   Construction

The counterpart of the production engineer in manufacturing is the construction engineer in the building industry (see Fig. 1.12). When an organization bids on a competitive construction project, the construction engineer begins the process by estimating material, labor, and overhead costs. If the bid is successful, a construction engineer assumes the responsibility of coordinating the project. On large projects, a team of construction engineers may supervise the individual segments of construction such as mechanical (plumbing), electrical (lighting), and civil (building). In addition to a strong background in engineering fundamentals, the construction engineer needs on-the-job experience and an understanding of labor relations.

1.3.6   Operations

Up to this point, discussion has centered around the results of engineering efforts to discover, develop, design, and produce products that are of benefit to humans. For such work engineers obviously must have offices, laboratories, and production facilities in which to accomplish it. The major responsibility for supplying such facilities falls on the operations engineer (see Fig. 1.13). Sometimes called a plant engineer, this individual selects sites for facilities, specifies the layout for all facets of the operation, and selects the fixed equipment for climate control, lighting, and communication. Once the facility is in operation, the plant engineer is responsible for maintenance and modifications as requirements demand. Because this phase of engineering comes under the

Figure 1.12

19
*The Functions of
the Engineer*

Numerous engineers from many disciplines are involved in the design and construction
of massive structures such as the dam shown.

Figure 1.13

Operations engineers help to lay out manufacturing facilities for optimum efficiency.

Figure 1.14

Sales engineers interface with people around the world using many forms of communications media.

economic category of overhead, the operations engineer must be very conscious of cost and keep up with new developments in equipment so that overhead is maintained at the lowest possible level. A knowledge of basic engineering, industrial engineering principles, economics, and law are prime educational requirements of the operations engineer.

1.3.7   Sales

In many respects all engineers are involved in selling. To the research, development, design, production, construction, and operations engineer, selling means convincing management that money should be allocated for development of particular concepts or expansion of facilities. This is, in essence, selling one's own ideas. Sales engineering, however, means finding or creating a market for a product. The complexity of today's products requires an individual who is thoroughly familiar with materials in and operational procedures for consumer products to demonstrate to the consumer in layperson's terms how the products can be of benefit. The sales engineer is thus the liaison between the company and the consumer, a very important means of influencing a company's reputation. An engineering background plus a sincere interest in people and a desire to be helpful are the primary attributes of a sales

Figure 1.15

21

*The Functions of
the Engineer*

Managers work long hours and often feel that they are buried in paperwork.

engineer. The sales engineer usually spends a great deal of time in the plant learning about the product to be sold. After a customer purchases a product, the sales engineer is responsible for coordinating service and maintaining customer satisfaction. As important as sales engineering is to a company, it still has not received the interest from new engineering graduates that other engineering functions have. See Fig. 1.14.

1.3.8 Management

Traditionally management has consisted of individuals who are trained in business and groomed to assume positions leading to the top of the corporate ladder. However, with the influx of scientific and technological data being used in business plans and decisions, there has been a need for people in management with knowledge and experience in engineering and science. Recent trends indicate that a growing percentage of management positions are being assumed by engineers and scientists. Inasmuch as one of the principal functions of management is to use company facilities to produce an economically feasible product, and decisions often must be made that may affect thousands of people and involve millions of dollars over periods of several years, a balanced education of engineering or science and business seems to produce the best managerial potential.

Figure 1.16

Consulting engineers designed and supervised the construction of this nuclear generating plant.

1.3.9   Consulting

For someone interested in self-employment, a consulting position may be an attractive one (see Fig. 1.16). Consulting engineers operate alone or in partnership furnishing specialized help to clients who request it. Of course, as in any business, risks must be taken. Moreover, as in all engineering disciplines, a sense of integrity and a knack for correct engineering judgment are primary necessities in consulting.

A consulting engineer must possess a professional engineer's license before beginning practice. Consultants usually spend many years in a specific area before going on their own. A successful consulting engineer maintains a business primarily by being able to solve unique problems for which other companies have neither the time nor capacity. In many cases large consulting firms maintain a staff of engineers of diverse backgrounds so that a wide range of engineering problems can be contracted.

1.3.10   Teaching

Individuals interested in a career that involves helping others to become engineers will find teaching very rewarding (see Fig. 1.17). The engineering

Figure 1.17

23

*The Engineering
Disciplines*

Teaching is a satisfying and rewarding profession.

teacher must possess an ability to communicate abstract principles and engineering experiences in a manner that young people can understand and appreciate. By merely following general guidelines, teachers are usually free to develop their own method of teaching and means of evaluating its effectiveness. In addition to teaching, the engineering educator also can become involved in student advising and research.

Engineering teachers today must have a mastery of fundamental engineering and science principles and a knowledge of applications. Customarily, they must obtain an advanced degree in order to improve their understanding of basic principles, to perform research in a specialized area, and perhaps to gain teaching experience on a part-time basis.

The emphasis in the classrooms today is moving from teaching to learning. Methods of presenting material and involving students in the learning process to meet designed outcomes are following sound educational principles developed by our education colleagues. You as a student will benefit greatly by these learning processes which empower you to take control of your education through teamwork, active participation, and hands-on learning.

If you are interested in a teaching career in engineering or engineering technology, you should observe your teachers carefully as you pursue your degrees. Note how they approach the teaching process, the methodologies they use to stimulate learning, and their evaluation methods. Your initial teaching methods likely will be based on the best methods you observe as a student.

1.4 The Engineering Disciplines

There are over 20 specific disciplines of engineering that can be pursued for the baccalaureate degree. The opportunities to work in any of these areas are

Figure 1.18

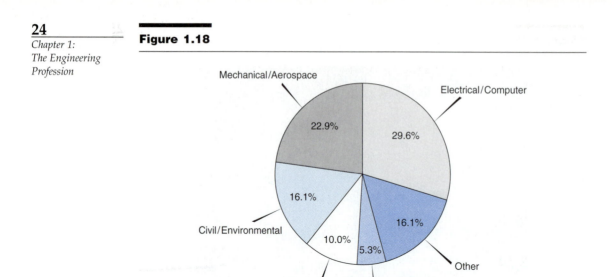

Mechanical/Aerospace

Electrical/Computer

22.9%

29.6%

16.1%

Civil/Environmental

16.1%

10.0%

5.3%

Other

Chemical/Petroleum

Industrial/Manufacturing/
Management

Engineering degrees by discipline. Total degrees awarded was 59 134. (*ASEE Profiles of Engineering & Technology Colleges, 1999 Edition*)

numerous. Most engineering colleges offer some combination of the disciplines, primarily as 4-year programs. In some schools two or more disciplines, such as industrial, management, and manufacturing engineering, are combined within one department which may offer separate degrees, or include one discipline as a specialty within another discipline. In this case a degree in the area of specialty is not offered. Other combinations of engineering disciplines include civil/construction/environmental, mechanical/aerospace, and electrical/computer.

Figure 1.18 gives a breakdown of the number of engineering degrees in six categories for 1999. Note that each category represents combined disciplines and does not provide information about a specific discipline within that category. The "other" category includes, among others, agricultural, biomedical, ceramic, materials, metallurgical, mining, nuclear, safety, and ocean engineering.

Seven of the individual disciplines will be discussed in this section. Engineering disciplines which pique your interest may be investigated in more detail by contacting the appropriate department, checking the library at your institution, and searching the World Wide Web.

1.4.1 Aerospace Engineering

Aerospace engineers study flight of all types of vehicles in all environments. They design, develop, and test aircraft, missiles, space vehicles, helicopters, hydrofoils, ships, and submerging ocean vehicles. The particular areas of specialty include aerodynamics, propulsion, orbital mechanics, stability and control, structures, design, and testing.

Figure 1.19

Many aerospace engineers work in avionics and cockpit design for new aircraft and upgrading of older aircraft.

Aerodynamics is the study of the effects of moving a vehicle through the earth's atmosphere. The air produces forces that have both a positive effect on a properly designed vehicle (lift) and a negative effect (drag). In addition, at very high speeds the air generates heat on the vehicle which must be dissipated to protect crews, passengers, and cargo. Aerospace engineering students learn to determine such things as optimum wing and body shapes, vehicle performance, and environmental impact.

The operation and construction of turboprops, turbo and fan jets, rockets, ram and pulse jets, and nuclear and ion propulsion are part of the aerospace engineering student's study of propulsion. Such constraints as efficiency, noise levels, and flight distance enter into the selection of a propulsion system for a flight vehicle.

The aerospace engineer develops plans for interplanetary missions based on a knowledge of orbital mechanics. The problems encountered include determination of trajectories, stabilization, rendezvous with other vehicles, changes in orbit, and interception.

Stability and control involves the study of techniques for maintaining stability and establishing control of vehicles operating in the atmosphere or in space. Automatic control systems for autopilots and unmanned vehicles are part of the study of stability and control.

The study of structures is primarily involved with thin-shelled, flexible structures that can withstand high stresses and extreme temperature ranges.

The structural engineer works closely with the aerodynamics engineer to determine the geometry of wings, fuselages, and control surfaces. The study of structures also involves thick-shelled structures that must withstand extreme pressures at ocean depths and lightweight composite structural materials for high-performance vehicles.

The aerospace design engineer combines all the aspects of aerodynamics, propulsion, orbital mechanics, stability and control, and structures into the optimum vehicle. Design engineers work in a team and must learn to compromise in order to determine the best design satisfying all criteria and constraints.

The final proofing of a design involves the physical testing of a prototype. Aerospace test engineers learn to use testing devices such as wind tunnels, lasers, strain gauges, and data acquisition systems. The testing takes place in structural laboratories, in propulsion facilities, and in the flight medium with the actual vehicle.

1.4.2   Chemical Engineering

Chemical engineers deal with the chemical and physical principles that allow us to maintain a suitable environment. They create, design, and operate processes that produce useful materials, including fuels, plastics, structural materials, food products, health products, fibers, and fertilizers. As our natural resources become scarce, chemical engineers are creating substitutes or finding ways to extend our remaining resources.

The chemical engineer, in the development of new products, in designing processes, and in operating plants, may work in a laboratory, pilot plant, or full-scale plant. In the laboratory the chemical engineer searches for new products and materials that benefit humankind and the environment. This laboratory work would be classified as research engineering.

In a pilot plant the chemical engineer is trying to determine the feasibility of carrying on a process on a large scale. There is a great deal of difference between a process working in a test tube in the laboratory and a process working in a production facility. The pilot plant is constructed to develop the necessary unit operations to carry out the process. Unit operations are fundamental chemical and physical processes that are uniquely combined by the chemical engineer to produce the desired product. A unit operation may involve separation of components by mechanical means, such as filtering, settling, and floating. Separation also may take place by changing the form of a component—for example, through evaporation, absorption, or crystallization. Unit operations also involve chemical reactions such as oxidation and reduction. Certain chemical processes require the addition or removal of heat or the transfer of mass. The chemical engineer thus works with heat exchanges, furnaces, evaporators, condensers, and refrigeration units in developing large-scale processes.

In a full-scale plant the chemical engineer will continue to "fine tune" the unit operations to produce the optimum process based on the lowest cost. The day-by-day operations problems in a chemical plant, such as piping, storage, and material handling, are the responsibility of chemical engineers.

Figure 1.20

27
*The Engineering
Disciplines*

**Chemical engineers design processing plants for many of the products on which we
depend in our daily lives.**

1.4.3 Civil Engineering

Civil engineering is the oldest branch of the engineering profession. The
term "civil" was used to distinguish this field from military engineers.
Military engineers originated in Napoleon's army. The first engineers trained
in this country were military engineers at West Point. Civil engineering involves application of the laws, forces, and materials of nature to the design,
construction, operation, and maintenance of facilities that serve our needs
in an efficient, economical manner. Civil engineers work for consulting firms
engaged in private practice, for manufacturing firms, and for federal, state,
and local governments. Because of the nature of their work, civil engineers
assume a great deal of responsibility, which means that professional registration is an important goal for the civil engineer who is beginning practice.

Figure 1.21

Civil engineers had a major role in the design of the route of this cross-country pipeline. The terrain had to be prepared to support the pipeline structure and care had to be taken to protect the local environment.

The specialties within civil engineering include structures, transportation, sanitary and water resources (environmental), geotechnical, surveying, and construction.

Structural engineers design bridges, buildings, dams, tunnels, and supporting structures. The designs include consideration of mass, winds, temperature extremes, and other natural phenomena such as earthquakes. Civil engineers with a strong structural background often are found in aerospace and manufacturing firms, playing an integral role in the design of vehicular structures.

Civil engineers in transportation plan, design, construct, operate, and maintain facilities that move people and goods throughout the world. For example, they make the decisions on where a freeway system should be located and describe the economic impact of the system on the affected public. They plan for growth of residential and industrial sectors of the nation. The modern rapid transit systems are another example of the solution to a public need that is satisfied by transportation engineers.

Sanitary engineers are concerned with maintaining a healthful environment by proper treatment and distribution of water, treatment of wastewater, and control of all forms of pollution. The water resources engineer specializes in the evaluation of potential sources of new water for increasing or shifting populations, irrigation, and industrial needs.

Before any structure can be erected, a careful study of the soil, rock, and groundwater conditions must be undertaken to ensure stability. In addition to these studies, the geotechnical engineer analyzes building materials, such as sand, gravel, and cement, to determine proper consistency for concrete and other products.

Surveying engineers develop maps for any type of engineering project. For example, if a road is to be built through a mountain range, the surveyors will determine the exact route and develop the topographical survey, which is then used by the transportation engineer to lay out the roadway.

Construction engineering is a significant portion of civil engineering, and many engineering colleges offer a separate degree in this area. Generally construction engineers will work outside at the actual construction site. They become involved with the initial estimating of construction costs for surveying, excavation, and construction. They will supervise the construction, start-up, and initial operation of the facility until the client is ready to assume operational responsibility. Construction engineers work around the world on many construction projects such as highways, skyscrapers, and power plants.

1.4.4   Electrical/Computer Engineering

Electrical/computer engineering is the largest branch of engineering, representing about 30 percent of the graduates entering the engineering profession. Because of the rapid advances in technology associated with electronics and computers, this branch of engineering also is the fastest growing.

The areas of specialty include communications, power, electronics, measurement, and computers.

We depend almost every minute of our lives on communication equipment developed by electrical engineers. Telephones, television, radio, and radar are common communications devices that we often take for granted. Our national defense system depends heavily on the communications engineer and on the hardware used for our early warning and detection systems.

The power engineer is responsible for producing and distributing the electricity demanded by residential, business, and industrial users throughout the world. The production of electricity requires a generating source such as fossil fuels, nuclear reactions, or hydroelectric dams. The power engineer may be involved with research and development of alternative generation sources, such as sun, wind, and fuel cells. Transmission of electricity involves conductors and insulating materials. On the receiving end, appliances are designed by power engineers to be highly efficient in order to reduce both electrical demand and costs.

The area of electronics is the fastest growing specialty in electrical engineering. The development of solid-state circuits (functional electronic circuits manufactured as one part rather than wired together) has produced high reliability in electronic devices. Microelectronics has revolutionized the computer industry and electronic controls. Circuit components that are much smaller than one micrometer in width enable reduced costs and higher electronic speeds to be attained in circuitry. The microprocessor, the principal component of a digital computer, is a major result of solid-state circuitry and microelectronics technology. The home computer, cellular telephones, automobile control

Figure 1.22

The computer is an indispensable tool for the engineer.

systems, and a multitude of electrical application devices conceived, designed, and produced by electronic engineers have greatly improved our standard of living.

Great strides have been made in the control and measurement of phenomena that occur in all types of processes. Physical quantities such as temperature, flow rate, stress, voltage, and acceleration are detected and displayed rapidly and accurately for optimal control of processes. In some cases the data must be sensed at a remote location and accurately transmitted long distances to receiving stations. The determination of radiation levels is an example of the electrical process called *telemetry.*

The impact of microelectronics on the computer industry has created a multibillion dollar annual business that in turn has enhanced all other industries. The design, construction, and operation of computer systems is the task of computer engineers. This specialty within electrical engineering in many schools has become a separate degree program. Computer engineers deal with both hardware and software problems in the design and application of computer systems. The areas of application include research, education, design engineering, scheduling, accounting, control of manufacturing operations, process control, and home computing needs. No single development in history has had as great an impact on our lives in such a short time span as has the computer.

1.4.5 Environmental Engineering

Environmental engineering deals with the appropriate use of our natural resources and the protection of our environment. For the most part environmental engineering curricula are relatively new and in many instances reside as a specialty within other disciplines, such as civil (see Sec. 1.4.3), chemical, and agricultural engineering.

The construction, operation, and maintenance of the facilities in which we live and work have a significant impact on the environment. Environmental

Figure 1.23

31
*The Engineering
Disciplines*

When engineers design a new product or system, such as an offshore drilling rig, the design must minimize the impact on the environment.

engineers with a civil engineering background are instrumental in the design of water and wastewater treatment plants, facilities that resist natural disasters such as earthquakes and floods, and facilities that use no hazardous or toxic materials. The design and layout of large cities and urban areas must include protective measures for the disturbed environment.

Environmental engineers with a chemical engineering background are interested in air and water quality which is affected by many by-products of chemical and biological processes. Products which are slow to biodegrade are studied for recycling possibilities. Other products which may contaminate or be hazardous are being studied to develop either better storage or replacement products that are less dangerous to the environment.

With an agricultural engineering background, environmental engineers study air and water quality which is affected by animal production facilities, chemical runoff from agricultural fertilizers, and weed control chemicals. As we become more environmentally conscious, the demand for designs, processes, and structures which protect the environment will create an increasing demand for environmental engineers. They will provide the leadership for protecting our resources and environment for generations to come.

1.4.6 Industrial Engineering

Industrial engineering covers a broad spectrum of activities in organizations of all sizes. The principal efforts of industrial engineers are directed to the

Figure 1.24

Industrial engineers design the assembly lines for production of products such as automobiles and trucks.

design of production systems for goods and services. As an example, consider the procedures and processes necessary to produce and market a power lawn mower. When the design of the lawn mower is complete, industrial engineers establish the manufacturing sequence from the point of bringing the materials to the manufacturing center to the final step of shipping the assembled lawn mowers to the marketing agencies. Industrial engineers develop a production schedule, oversee the ordering of standard parts (e.g., engines, wheels, and bolts), develop a plant layout (assembly line) for production of nonstandard parts (frame, height adjustment mechanism), and perform a cost analysis for all phases of production.

As production is ongoing, industrial engineers will perform various studies, called *time and methods studies,* which assist in optimizing the handling of material, the shop processes, and the overall assembly line. In a large organization industrial engineers will likely specialize in one of the many areas involved in the operation of a plant. In a smaller organization industrial engineers are likely to be involved in all the plant activities. Because of their general study in many areas of engineering and their knowledge of the overall plant operations, industrial engineers are frequent choices for promotion into management-level positions.

The study of human factors is an important area of industrial engineering. In product design, for example, industrial engineers involved in fashioning automobile interiors study the comfort and fatigue factors of seats and instrumentation. And in the factory they develop training programs for operators and supervisors of new machinery or for new assembly-line operators.

With the rapid development of computer-aided manufacturing (CAM) techniques and of computer-integrated manufacturing (CIM), the industrial engineer will continue to play a large role in the factories of the future. The industrial engineer of the future also will be involved in training and retraining the labor force to work productively in a high-technology environment.

1.4.7   Mechanical Engineering

Mechanical engineering originated when people first began to use levers, ropes, and pulleys to multiply their own strength and to use wind and falling water as a source of energy. Today mechanical engineers are involved with all forms of energy utilization and conversion, machines, manufacturing materials and processes, and engines.

Mechanical engineers utilize energy in many ways for our benefit. Refrigeration systems keep perishable goods fresh for long periods of time, air condition your homes and offices, and aid in various forms of chemical processing. Heating and ventilating systems keep us comfortable when the environment around us changes with the seasons. Ventilating systems help to keep the air around us breathable by removing undesirable fumes. Mechanical engineers analyze heat transfer from one object to another and design heat exchangers to effect a desirable heat transfer.

The energy crisis of the 1970s brought to focus a need for new sources of energy as well as new and improved methods of energy conversion. Mechanical engineers are involved in research in solar, geothermal, and wind energy sources, along with research to increase the efficiency of producing electricity from fossil fuel, hydroelectric, and nuclear sources.

Machines and mechanisms used in all forms of manufacturing and transportation are designed and developed by mechanical engineers. Automobiles, airplanes, and trains combine a source of power and a machine to provide transportation. Tractors, combines, and other implements aid the agricultural community. Automated machinery and robotics are rapidly growing areas for mechanical engineers. Lathes, milling machines, grinders, and drills assist in the manufacture of goods. Sorting devices, typewriters, staplers, and mechanical pencils are part of the office environment. Machine design requires a strong mechanical engineering background and a vivid imagination.

In order to drive the machines, a source of power is needed. The mechanical engineer is involved with the generation of electricity by converting chemical energy in fuels to thermal energy in the form of steam, then to mechanical energy through a turbine to drive the electric generator. Internal combustion devices such as gasoline, turbine, and diesel engines are designed for use in all

Figure 1.25

Final touches are made to a large backhoe designed by mechanical engineers.

areas of transportation. The mechanical engineer studies engine cycles, fuel-requirements, ignition performance, power output, cooling systems, engine geometry, and lubrication in order to develop high-performance, low-energy-consuming engines.

The engines and machines designed by mechanical engineers require many types of materials for construction. The tools that are needed to process the raw material for other machines must be designed. For example, a very strong material is needed for a drill bit that must cut a hole in a steel plate. If the tool is made from steel, it must be a higher quality steel than that found in the plate. Methods of heat treating, tempering, and other metallurgical processes are applied by the mechanical engineer.

Manufacturing processes, such as electric-discharge machining, laser cutting, and modern welding methods are used by mechanical engineers in the development of improved products. Mechanical engineers also are involved in the testing of new materials and products such as composites.

1.5 Total Quality in Engineering

Now that you have some idea of the functions of an engineer in an industrial or educational climate and the many engineering disciplines that are available for study at an engineering school, we can introduce you to the industrial environment through the concept of total quality. An industry or business exists to make a profit for its owners, whether the ownership is private or public. To maximize the performance (profit) of the industry, management and workers must come together in an optimum manner to compete effectively in the marketplace. As an engineer in an organization, you are expected to function in a manner that focuses on the goals of the organization and that creates harmonious relationships among all those with whom you interact with in the course of your work.

As a beginning engineering student who is busy with a variety of coursework and extracurricular activities you may not have the time or opportunity during your first year in engineering to experience a complete total quality process. In fact, our intent in this introductory chapter is not to provide a complete detailed description of the entire process with sufficient information together with the necessary training to carry out a full-scale project. Rather, the objective of this discussion is to provide an abbreviated overview that accomplishes these purposes:

1. To present a brief review of the background and history of total quality.
2. To develop an appreciation for the importance of total quality and its role in making a business or industry internationally competitive.
3. To provide an introduction to many definitions, unique terminology, and elements that apply to total quality.
4. To explore two important aspects of total quality that relate and are critically important in engineering: problem-solving models and the team process.

We believe that with this background you will understand better how total quality is destined to play an important role in your life as both an engineering student and eventually a professional engineer.

Total quality (TQ), total quality management (TQM), and continuous quality improvement (CQI) are three of the many titles given to the central concept of organizational behavior and management. In this chapter we will use the term "total quality," or TQ, but it is the philosophy, not the title, that is fundamental.

As a student in engineering there are several ways that this idea will impact on your immediate life. While you read this material and begin the process of understanding TQ, it is likely that this subject area has been or is currently being integrated into your undergraduate curricular activities. This introduction to TQ could be followed with additional coursework and application or team involvement in a variety of other areas. A second way it may affect you is by virtue of the fact that you are a "customer" of some unit or agency that is striving to provide significantly improved customer satisfaction. For example, you have probably completed many customer surveys asking about your satisfaction with products or services provided to you. A third way you could be affected would be to do some additional individual reading and to apply TQ principles to your personal life, habits, and actions. One such reference for consideration would be *Quality Is Personal* by H. V. Roberts and B. F. Sergesketter.[1]

---

[1] H. V. Roberts and B. F. Sergesketter, *Quality Is Personal*, New York: The Free Press, 1993.

One ancillary objective or outcome of this introductory material is to help you understand that TQ is a subject that you can read easily and understand. It is not a complex or difficult subject to comprehend. If, however, you intend to practice TQ effectively, a totally different approach is required. A commitment to TQ demands a modification in the manner in which you approach problem solving as an engineer; it involves a completely new and different way of thinking—a philosophically different attitude toward quality. Many opportunities to learn about and practice TQ principles will be available to you in your college studies and extracurricular activities.

1.6 Background and History of TQ

The information summarized in Table 1.1 contains selected writings and authors who have made significant contributions in the last 100 years. Other references are contained in footnotes throughout this discussion.

The amount of written material about TQ has significantly expanded in the past few years. From 1990 to the present there have been many books and numerous papers written on the subject. Only two are referenced here because they apply directly to the impact of TQ in higher education:

1992 *On Q: Causing Quality in Higher Education* Daniel Seymour
1995 *Malcolm Baldrige National Quality Award*

In the late 1970s and early 1980s international competition for market share was becoming a way of life: The United States was being outperformed by foreign companies in too many areas. Perhaps a simple way to illustrate this is by a brief example.

In the 1970s Japanese companies began to take market shares from many U.S. companies. David Kearn, who was at that time chairman of Xerox, tells it like this:[2]

> Whereas most American corporations were advancing 2 or 3 percent a year in productivity, we were achieving gains of 7 or 8 percent. But despite these gains, the Japanese continued to price their products substantially below us. We kept wondering: How were they doing it? Our team went over everything in a thorough manner. It examined all the ingredients of cost: turnover, design time, engineering changes, manufacturing defects, overhead ratios, inventory, how many people worked for a foreman, and so forth. When the team completed its calibrations, we were quite shocked.

From this same reference another Xerox executive, Frank Pipp, remembers the results of the review as follows:

> Absolutely nauseating, it wasn't a case of being out in left field, we weren't even playing the same game.

The Xerox experience was, in fact, an example of what was happening with great frequency in the 1970s and 1980s. For corporate America this was

---

[2]David T. Kearnes and David A. Nadler, *Prophets in the Dark: How Xerox Reinvented Itself and Beat Back the Japanese*, New York: Harper Business, 1992.

Table 1.1 The Origin and Evolution of Total Quality

Date	Subject Areas	Author/Reference
1911	Principles of Scientific Research	Frederick W. Taylor
1924	Control Charts	W. A. Shewhart
1927	Human Relations Influence on Productivity	Hawthorne Research
1947	Administrative Behavior: Systems Theory	Herbert Simon
1950	Japanese Quality Movement	W. Edwards Deming
1951	Change Model	Kurt Lewin
1954	Quality Control Management in Japan	J. M. Juran
1962	Quality Circles	Kaoru Ishikawa
1962	Paradigms	Thomas S. Kuhn
1966	Open Systems Theory	Herman Kahn
1968	Cost of Quality	Armand Feigenbaum
1979	*Quality Is Free*	Philip Crosby
1980	If Japan Can . . . Why Can't We?	NBC Documentary
1982	*In Search of Excellence*	Tom Peters
1982	*Out of the Crisis*	W. Edwards Deming
1985	Organizational Culture	Edgar Schein
1985	Leadership	Warren Bennis and Burt Nanus
1985	Strategic Thinking and Visioning	John W. Zimmerman
1986	*The Transformational Leader*	Noel M. Tichy
1987	Empowerment	Peter Block
1988	National Quality Award	Malcolm Baldrige
1990	International Quality Requirements	ISO 1900
1990	*Quality or Else*	Lloyd Dobyns
1990	*The Fifth Discipline*	Peter M. Senge

a wake-up call. The manufacturing and service industries, long thought to be the envy of the entire planet, were no longer number one and in many cases were not even competitive.

American industry became justifiably alarmed by the lack of competitiveness of the United States in the global marketplace. In the eight critical industries shown in Fig. 1.26, the United States had a positive balance of trade in only two at the time the data was collected. The other six industries represent areas where the United States was once dominant, but subsequently foreign competition has eroded the U.S. position.

To understand better how this could happen, let us take a brief look at our history and how certain events allowed this country to lose global competitiveness in a number of areas.

Frederick Taylor (1856–1917)

Taylor is generally considered to be the father of modern time study in this country. His thoughts and ideas suggested that the work of each employee should be planned by management and that each worker should receive complete written instructions describing the task to be done in detail. Each job was to have a standard time for completion, which was to be fixed after time studies had been done by experts. In the timing process, Taylor advocated breaking up the work assignment into small dimensions of effort known as "elements." These were timed individually and their collective values were used to determine the allowed time of a given task.

It is important to note that in the late 1800s this approach was needed because the workforce was largely untrained and uneducated. Taylor asked the following specific question: "What is the best way to do a job?" He did not accept opinion; he demanded facts and evidence. As a mechanical engineer, he was the first to apply the principle of the scientific approach to the application of improving jobs. At that time he stated, "Any change the worker makes to the plan is fatal to success." The Taylor approach was a system that separated planning from execution but also tended to separate management from workers. Its major premise was that the supervisor and the worker lacked the education needed to plan how work should be done; hence planning was turned over to management and engineering.

Today this approach seems contrary to the TQ concepts; however, we need to remember that in the late 1800s and early 1900s this approach significantly increased efficiency and productivity.

W. Edwards Deming

As a young man Deming worked for Walter Shewhart, who was the creator of statistical process control and the inventor of the control chart. In the 1940s Deming worked for the U.S. Bureau of the Census where he applied quality principles and taught courses in statistical process control. Therefore his early

Figure 1.26

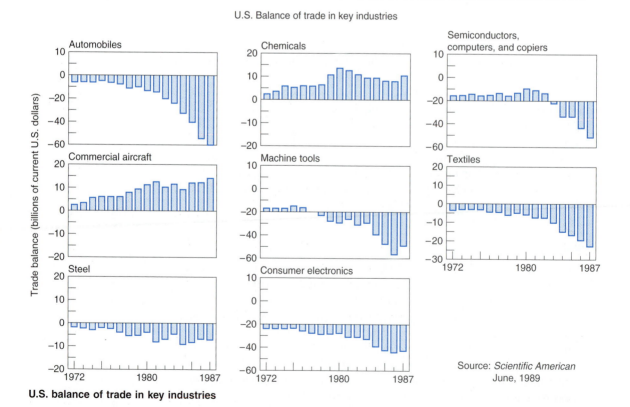

U.S. Balance of trade in key industries

U.S. balance of trade in key industries

Source: *Scientific American*
June, 1989

Table 1.2   W. Edwards Deming's 14 Points

1. Create constancy of purpose for the improvement of product and service.
2. Adopt the new philosophy. We are in a new economic age. Poor workmanship and sullen service are no longer tolerable. Stay in business and provide jobs through innovation, research, and constant improvement.
3. Use modern science. Cease dependence on inspection to achieve quality. Quality comes not from inspection but from improvement of the process. Build quality into the product/process.
4. End the practice of awarding business on the basis of price tag. Keep total costs in mind, not partial costs. Seek the best quality, not the cheapest price.
5. Improve constantly and forever the system of production and service, to improve quality and productivity, and thus constantly decrease costs. Find problems. Management must work continually on the system.
6. Never stop training on the job.
7. Improve supervision of workers and managers. Institute leadership. The aim of leadership should be to help people, machines, and gadgets to do a better job.
8. Drive out fear so everyone may work effectively for the organization.
9. Break down barriers between departments. An organization must work as a team.
10. Eliminate slogans, exhortations, and targets for workers asking for better quality and productivity.
11. Eliminate work standards that have only numerical quotas. Quotas take account only of numbers, not quality or methods. Eliminate management by objectives. Eliminate management by numbers and numerical goals. Substitute leadership.
12. Remove barriers that stand between workers and their right to pride in workmanship.
13. Institute a vigorous program of education and self-improvement.
14. Put everybody in the organization to work to accomplish the transformation. The transformation is everybody's job.

contribution was as a statistician. He taught the value of statistics as a tool for measuring the output of a process within a system. He believed that organizations are systems with processes. What managers do is to manipulate processes, consciously or unconsciously, skillfully or erratically, to get work done.

Deming had an enormous influence on the start and subsequent progress of the "quality revolution." His methods of measurement and process control resulted in significant improvement to the quality of American products prior to World War II. Following the war, the U.S. focus shifted from quality to quantity to meet postwar demands. Meanwhile Deming was invited to Japan to help rebuild their infrastructure. Deming, along with another American, Joseph M. Juran, was very influential in the 1950s, starting the quality improvement movement in Japan. Much of Deming's management philosophy is contained in what he calls his "14 points." Over the course of many years, through both practice and teaching, Deming modified and improved this list of 14 points. He believed these ideas to be an essential way of life for corporate America.

Deming's 14 points are listed in Table 1.2 because they are an important part of the TQ evolution. However, without considerable explanation they are difficult to understand totally. For you as a student they are included as an important reference.[3]

---

[3]W. Edwards Deming, (1986), *Out of the Crisis*, MIT Center for Advanced Engineering Study, Cambridge, MA.

Table 1.3 Philip B. Crosby's 14 Steps of Quality Improvement

1. ***Management commitment.***   Management must be committed. They must be personally committed and provide leadership by example, in other words, "Walk the talk."
2. ***Quality improvement teams.***   Collect the correct individuals into a quality improvement team. Must be properly engaged.
3. ***Quality measurements.***   Quality measurements and current status are necessary to determine where corrective action is necessary and to document actual improvement.
4. ***Cost of quality evaluation.***   The cost of quality is not an absolute performance measure, it is an indication of where corrective action will be profitable.
5. ***Quality awareness.***   The process of sharing with employees the measurements and results of the amount nonquality is costing. Getting supervisors and employees in the habit of talking positively about quality.
6. ***Corrective action.***   An opportunity for workers to contribute to the solution set.
7. ***Establish an ad hoc committee for zero defects.***   The idea is that things should be done correctly the first time. Defects cost, and zero defects provide a better product with less cost.
8. ***Supervisor training.***   Training in total quality concepts and process is critical. Everyone must clearly understand the ideas and philosophy.
9. ***Zero defects day (ZD).***   Establishment of ZD as a performance standard of the company should be done in one day. New company attitude.
10. ***Goal setting.***   Supervisors should ask employees to establish 30-, 60-, and 90-day goals.
11. ***Error cause removal.***   Describe any problem that keeps them from performing error-free work.
12. ***Recognition.***   Award programs are established to recognize those who meet their goals.
13. ***Quality councils.***   Quality professionals and the team chairs should be brought together to exchange, learn, and redefine the quality program.
14. ***Do it over again.***   A typical quality team project takes from 12 to 18 months to complete. During that length of time many company employees may be transferred, leave, and so on—so DO IT AGAIN.

Philip B. Crosby

Crosby is chief executive of PCA, Inc., a quality management consulting and teaching firm he founded. Previously he was corporate vice president and director of quality for the ITT Corporation, responsible for worldwide quality operations. He is best known as the creator of the "zero defects" concept. Crosby explains zero defects as follows:

> Some American companies were in the habit of spending 15–20 percent of every sales dollar on reworking, scrapping, repeating, inspecting, testing, and other quality-related costs. This results from the traditional idea that quality control happens only on the manufacturing line. The idea that "quality is free" is meant to suggest that whatever you save from the fact that "unquality" costs money therefore increases your profit.

Crosby also developed 14 foundation blocks that he believes must be followed.[4] He stated that each step is complex and that no step can be eliminated. His 14 steps of quality improvement are included in Table 1.3.

---

[4]Philip B. Crosby, *Quality Is Free*, McGraw-Hill Book Company, Copyright © 1979.

Total quality (TQ) is a process, and any process is made up of its own unique array of terminology and definitions. This section provides an overview of the essential concepts associated with TQ.

1.7.1   Definition

A simplistic definition of TQ could be stated as:

> Meeting or exceeding the expectations of the customer.

Other more comprehensive definitions follow.[5]

The Report of the Total Quality Forum

Total quality is a people-focused management system that aims at continual increase of customer satisfaction at continually lower real cost. TQ is a total system approach, an integral part of organizational high-level strategy. It expands horizontally across functions and departments, it involves all employees—top to bottom—and extends backwards and forwards to include the supply chain and the customer chain. TQ stresses learning and adaptation to continual change as keys to organizational success.

Thomas Johnson

The key feature of this paradigm is the idea that work is a process and that business is a system of processes that is aimed at exceeding customer expectations profitably.

Carla O'Dell

A process is defined as an activity that is definable, repeatable, measurable, and predictable. Process management is a methodology that attempts continually to increase effectiveness and efficiency of processes.

The process of identifying an organization's functions and breaking these functions into finer and finer subfunctions is called function analysis or function deployment.

Processes are definable, predictable, repeatable actions that can be flow-charted. Processes are particular actions that deliver functions. For example, a key function of every organization is "to understand markets and customers"; hence every organization needs a process by which this is accomplished. From industry to industry and organization to organization there are many different processes for delivering a given function.

Don Clausing

Management by fact applied by teams to improve processes (as perceived by customers) with high expectations.

A common expression often heard when discussing total quality is that it begins with education and ends with education. Learning the process and system whether as an individual or as part of a team takes much effort, time, and

---

[5]*A Report of the Total Quality Leadership Steering Committee and Working Councils,* November 1992, published by John K. Lowe Company, Cincinnati, OH.

commitment. This section introduces some of the elements and statistical tools used in the process.

1.7.2   Elements of Total Quality

Some of the critical elements of TQ that must be present or included in the process are discussed in the following sections.

Top-Level Management Commitment and Involvement

It is completely apparent to everyone if top management or the central administrative team of an organization is all "talk" and no "do", commitment must be real and visible. The late Jerry Junkins, former CEO of Texas Instruments, once stated that "Top management must be convinced. The idea that this concept of Total Quality is necessary for you but I'm already perfect does not work. Top administration must 'walk the talk.'"

Continuous Incremental Improvement

A statement overheard from an industrial CEO (to remain nameless) was as follows:

> "One of the problems that many organizations unfortunately face is that they cruise along in this state of uninformed euphoria and do not notice (or care) how out of touch or how bad things are becoming. One day the ax falls." When this is the case the resulting action normally means sweeping or radical change. No one likes or enjoys this outcome. Instead of experiencing the big changes, organizations should make hundreds of small improvements. People normally embrace and respect incremental change, particularly if employees are involved.

Focus on Process

A process is the steps that depict how the system achieves its objectives. To illustrate the concept of process, we often use flowcharts where a box represents a single step in that process. W. Edwards Deming discussed process as follows: "Every activity, every job is a process. A flow diagram of any process will divide the work into stages. The stages as a whole form a process. Each stage works with the next stage and with the preceding stage toward optimum accommodation."[6]

All organizations are composed of a multitude of processes (they can be stated and documented or unstated and undocumented) to get work done. Every organization that delivers a product or a service goes through a series of steps to produce something of value to a customer. This is why a focus on process is central. Organizational managers should carefully analyze the processes in their organization to understand how the work gets done.

The concept *kaizen* (improvement) is at the core of the way the Japanese manage their organizations. This philosophy has generated a process-oriented way of thinking, and they have developed strategies that assure continuous improvement through the involvement of people at all levels within the organizational hierarchy.

---

[6]Deming, *Out of the Crisis*, 1986.

Think of any organization as the place where countless tasks are accomplished. Perhaps it is putting addresses on letters, tightening a machine screw, or removing burrs from a shearing process.

A process can be defined as each of the individual tasks grouped into a sequence necessary for a unique outcome. Every organization has thousands of these processes. If you can improve one or many of these processes, you can improve the outcome, which means better quality. People who view work as a process understand how the quality of what comes out is a function of the quality that goes in. Thus there is the need to review the product received and to insist that suppliers provide quality input.

Many times the TQ team will try to tackle an entire system that produces a product. Systems are composed of many processes. The production of a product involves thousands of interrelated processes, and it may be necessary for the team to narrow or focus on a specific portion.

Employee Involvement

Employees in today's organizational workforce are educated, have job experience, are creative, and know their particular task better than others. They have ideas about how to improve the process if asked. Most employees are willing, perhaps eager, to share ideas. Management does not have all the answers. Management must work with employees to identify areas that need improvement, and then get the workforce involved in providing suggestions for improvement. Once the people are engaged in the process, they are more committed to its implementation (i.e., their ideas) and to its success.

The Customer

Defining exactly who is the customer often can be complex. For example, if you are within an organization looking at the entire system, the customer may be defined as the individual who purchases a product. Most quality teams within an organization are only looking at a single process in a series of events within the larger system. Who then is their customer? How does a given customer define quality? Is it defined the same by all customers?

Consider the suppliers of materials as people in the organization who precede you in a series of tasks and the customers as those who receive the product or service. If customers are the people who receive your work, only they can tell you what they expect in terms of quality.

Competitiveness

The late Kaoru Ishikawa, one of Japan's most distinguished quality gurus, believed that "95 percent of the quality problems in the workplace could be solved using what are known as the Seven Quality Control Tools." These seven tools establish the basis of the statistical quality control process which is a major component of TQ:

1. *Check sheet/checklist.* Makes it easy to collect and use data.
2. *Pareto diagram.* Helps to find the "vital few" causes that create the majority of the problems.
3. *Histogram.* Describes the manner of dispersion of data.
4. *Cause-and-effect diagram (Ishikawa diagram).* Arranges the cause-and-effect relationship.

5. **Stratification.** Helps to find differences between data.
6. **Scatter diagram.** Shows the correlation between two variables.
7. **Graph and control chart.** Helps to find a specific feature behind the data and to analyze and control the process.

Data

The TQ process and effective team results rely heavily on data. The eventual ability of the team to analyze and synthesize a solution requires a solid understanding of process together with an ability to visualize. These elements rely on a variety of tools.

Diagrams	**Charts**
Activity network diagram	Gantt chart
Affinity diagram	Control chart
Cause-and-effect diagram	Flowchart
Matrix diagram	Pareto chart
Scatter diagram	Radar chart
Tree diagram	Run chart

Application

As you have an opportunity to be selected as a TQ team member, either in school or after graduation, you will need to explore many of these tools learning their unique and special applications.

The Pareto Chart

The Pareto chart displays the relative importance of one item to another in a simple, visual format. Once the major problem has been identified and significant improvement realized, the team can pursue other problems in the quest for continuous quality improvement. The Pareto chart is based on the 20-to-80-percent rule; that is, 20 percent of the problem sources cause 80 percent of the problems. The key or fundamental reason for using a Pareto chart is to focus efforts of the team on the single problem that offers the greatest potential for improvement.

An example will illustrate how the Pareto chart works. The data in Table 1.4 were taken from an actual TQ team organized at Iowa State University to look

Table 1.4 Engineering Retention Survey Results

Problem Area	Frequency	Percent
Coursework related	160	38
Uncertainty as to "what is engineering?"	92	22
Motivation (attitude)	55	13
Personal problems	38	9
Poor precollege preparation	21	5
Poor instruction	13	3
Poor advising	8	2
Others	33	8
	420	100

Figure 1.27

45

Definitions and Tools

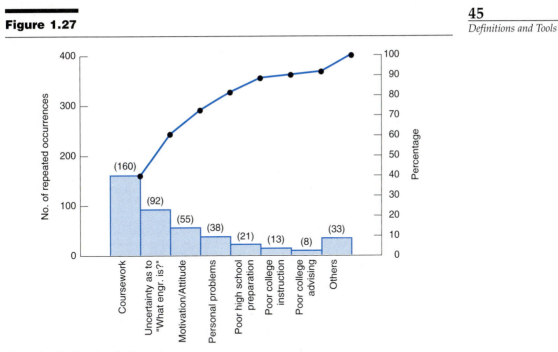

Example of a Pareto chart.

at the issue of freshman engineering retention. The team conducted personal interviews with 50 students and 10 faculty advisors.

Once the data are collected, the next step is to create the Pareto chart. Each of the categories is listed in descending order from left to right on the horizontal axis (see Fig. 1.27). Next establish bars above each problem area corresponding in height to the frequency of occurrence on the y axes. Last, construct the cumulative percentage line showing the portion of the total that each area represents.

Again, the tallest bar graphically illustrates the single largest contribution to the overall problem. It may be that the team can have absolutely zero impact on that particular problem area and will select the second or third problem area for action. The chart, however, guides the team toward solutions that will have a major impact on the problem.

Cause-and-Effect Diagram

This tool, often called the Ishikawa fishbone diagram, allows a team to identify and graphically display, in increasing detail, many of the possible "causes" related to a specific "effect" or condition.

To continue the retention example, the number one problem area identified was coursework. First, the team modified this concise problem statement to allow as much flexibility as possible. They agreed that the following statement summarized the problem area: "Students are not successful in first-year coursework."

What are the causes that produce this effect? The effect is placed in a box on the right-hand side of the diagram (see Fig. 1.28). Major causes are then

Figure 1.28

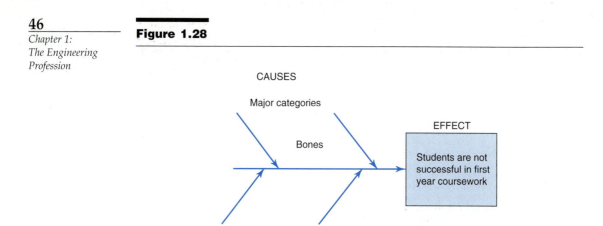

Cause-and-effect, or fishbone, diagram structure.

developed and placed on the basic bone of the chart. There is no set number of categories.

Figure 1.29 is the result of a "brainstorming" session the team members conducted using data they had collected as well as their personal experiences.

The Team

Why would we elect to form a team instead of having a talented individual solve a particular problem? Individuals can do great things, but rarely does a single person have the knowledge and experience needed to consider all elements in a process. Major gains in quality and productivity can be derived from a group of people pooling their talents and skills to tackle a complex problem.

Figure 1.29

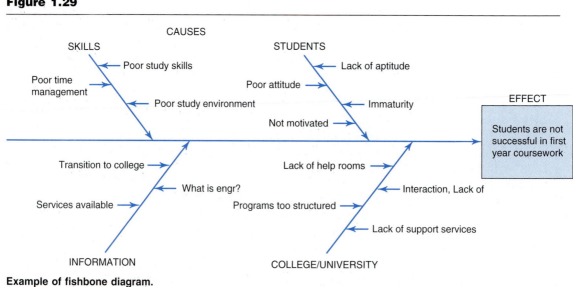

Example of fishbone diagram.

As with most issues, there are advantages and disadvantages to team versus individual effort. One disadvantage is that teams require time, patience, and the learning of new skills to deal effectively in a team environment. However there are advantages once a team is working together effectively, for example, the mutual support received from other team members while working on a project that may take considerable time and effort.

Scientific Approach

The central theme of quality improvement methods can be summed up in two words: scientific approach. Put simply, this means that there are systematic ways for individuals and teams to learn about a process; that is, making decisions based on facts derived from data rather than from a hunch, and searching for root causes of problems. A famous TQ expert stated during a TQ meeting: "In God we trust, others must have data."

Culture Change

Culture change involves a modification in the fundamental way we think and act, and it does not happen easily or quickly. The TQ process demands a change of culture within each individual and therefore throughout an organization: It means that the organization collectively thinks differently, reacts spontaneously; that the organization supports a totally different approach to the solution of problems, one that focuses on quality; and that every person is committed to quality. All events consist of processes, and all processes can be improved. "Zero defects" (ZD) really means, do it right the first time. Total quality means understanding customer needs and exceeding those expectations, never accepting conformity and always understanding that continuous incremental improvement is possible, and all employees must be involved because they are the organization and must base all decisions on fact. When every employee is involved, the organizational environment is positive and healthy, the product or service is continually improving, and the customer is delighted; thus the organization has experienced the TQ concept.

1.8 Problem Solving and the Team in TQ

This section introduces two of the many problem-solving models available for use during the TQ process, but the material presented will focus primarily on the team and not on the model. We will give ideas about working in a group environment and include a look at personal style inventory.

1.8.1 Problem-Solving Models

A clearly defined problem-solving model in the TQ arena accomplishes three things:

1. It helps all team members to understand the current situation.
2. It provides a detailed assessment of customer requirements.
3. It drives solutions by facts. The team relies on data analysis and measurement to justify selection of solutions.

Peter R. Scholtes in *The Team Handbook*[7] has developed a series of strategies or steps to follow. By using these strategies, you can design a plan. Scholtes outlines the following 14 improvement strategies :

A. Strategies of the Scientific Approach
1. Collect meaningful data.
2. Identify root causes of problems.
3. Develop appropriate solutions.
4. Plan and make changes.

B. Strategies for Identifying Improvement Needs
5. Identify customer needs and concerns.
6. Study the use of time.
7. Localize recurring problems.

C. Strategies for Improving a Process
8. Describe a process.
9. Develop a standard process.
10. Error-proof a process.
11. Streamline a process.
12. Reduce sources of variation.
13. Bring a process under statistical control.
14. Improve the design of a product or process.

Although these steps appear quite simple, most of them are more complex than they first appear and frequently lead to other meaningful events. For example, Strategy 3, Develop Appropriate Solutions, involves a seven-step process that is, in fact, very close to the traditional design process that engineers follow in the solution of open-ended problems.

Strategy 3: Develop Appropriate Solutions	**Engineering Design Process**
1. Describe the need.	1. Identify the need.
2. Define goals and criteria.	2. Define the problem.
3. Identify constraints.	3. Search.
4. Generate alternatives.	4. Criteria.
5. Evaluate alternatives.	5. Constraints.
6. Select best solution.	6. Alternative solutions.
7. Follow up.	7. Analysis.
	8. Decision.
	9. Specification.
	10. Communication.

Open-ended problems can be divided into two broad categories: the problem and the solution. A problem is a situation or condition that needs to be changed or improved. Problem solving, on the other hand, is the action taken to change the existing situation. Interactive problem solving is a process that allows people to work together productively to address issues and to create opportunities. Most often the desired result is a consensus decision. Familiarity

---

[7]Peter R. Scholtes, *The Team Handbook*, Madison, WI, Joiner Associates Inc., 1992.

Figure 1.30

49
*Problem Solving and
the Team*

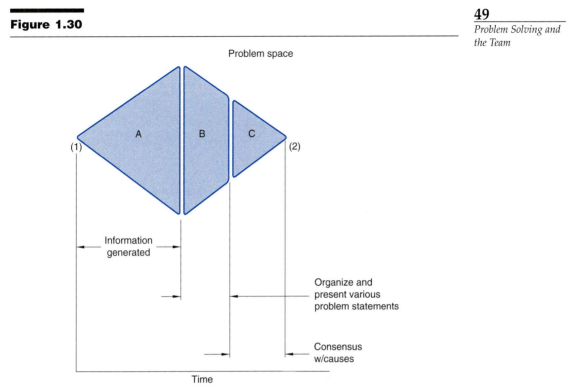

Visual diagram of problem space.

with open-ended problem-solving strategies is essential for individuals and teams to work well together, to resolve issues, and to make quality decisions.

Problem Space

In the beginning the team must develop a mutual understanding of all aspects of the problem. They must agree on a problem definition and identify root causes so that the eventual solutions can address these important causes rather than just symptoms. In other words, the team must agree on what the problem really is before beginning to work toward a solution. This will involve considerable discussion and a search for needed information. Proper identification of the problem is a critical step. The diamond shape shown in Fig. 1.30 represents three necessary phases that a team must move through as the problem is completely defined. Starting at location (1) the team must learn as much as possible about the problem. Phase A represents a large increase in the amount of information. Phase B begins the process of sorting and organizing information. It introduces a range of problem statements. After considerable discussion agreement by the team is reached in phase C. Once the group has collectively decided on the best problem statement, location (2), it is ready to move into solution space.

Solution Space

Figure 1.31 illustrates the process of solution generation. A number of techniques are available to assist a team during this important phase. One common

Figure 1.31

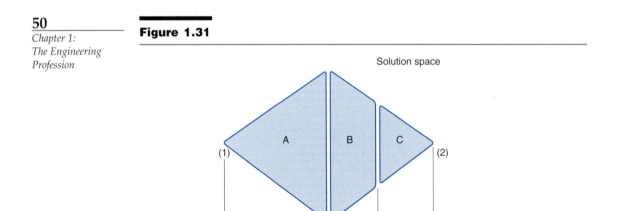

Visual diagram of solution space.

method is called "brainstorming." The diagram depicts the process of expanding solution space to include as many ideas as possible. Then the process of narrowing must take place. This involves both evaluation of ideas against a predetermined list of criteria and the process of decision making.

The important point to be understood is that there are many adaptations and variations of the scientific method of problem solving. Depending on the situation and what the goals or objectives of the team project happen to be, a problem-solving model can be or already has been developed to assist and direct this effort.

The undergraduate engineering educational process provides students with many fundamental engineering science courses that are essential to the professional engineer. However, the educational mission also includes practicing the process—that is, learning by application. Through repetition we develop optimum open-ended problem-solving skills.

A different process or series of steps is suggested by a problem-solving model developed at Oregon State University. Their approach to the problem-solving process suggests the following steps for the team:

1. Interview customers.
2. Select issues and develop performance measures.
3. Diagram the process.
4. Diagram cause-and-effect performance measures.
5. Collect and analyze data on causes.

6. Develop solutions.
7. Benchmark.
8. Select and implement solutions.
9. Measure results and refine.

Each of these 9 steps contains additional menus of items that need to be considered. For example, step 8 (select and implement solutions) recommends an additional process: plan-do-check-act. This process, called the P-D-C-A cycle, was used years ago by Walter Shewhart and is sometimes referred to as the Shewhart cycle.

Most of this terminology and many of these steps have been used by engineers for years. How and where they can be applied to TQ is an ongoing learning experience. A considerable amount of trial-and-error discovery as well as targeted research is currently under way in the TQ movement. Much of the work will be learning to improve current practice, but the single biggest change needed is not in learning or applying the process but in "behavior modification"—that is, changing the way we think.

Too many organizations and individuals currently operate with a mentality that believes if there is a problem, fix it and move on to something else; if no problem exists, there is no problem to fix. Today the attitude must be to strive constantly and continuously to improve processes even when no problem exists. If we are aware of problems, naturally start with the big problems but search continuously for better ways to improve all processes. Never be complacent or satisfied—always seek improvement.

1.8.2   The Team

You will spend a high percentage of both your professional and personal life interacting with people. Some of these interactions will be one-on-one, but many will involve groups of individuals who are either self-selected or members who have been assigned to committees or teams. For example, in the school of engineering you likely will be assigned as a member of a design team or you may be a team member in a collaborative learning situation. Design team membership is typically assigned by the instructor although occasionally team membership can be self-selecting. Design teams typically consist of four to six members and these individuals work together for the duration of the academic term. The team and its membership in the TQ process is more prescribed. The structure consists of a sponsor, a team leader, a facilitator, a recorder, and the individual members.

Sponsor

The sponsor is responsible for translating the "company" mission and vision into appropriate action. This individual would typically be a company manager, but in the academic environment it would be your instructor.

Team Leader

The team leader has responsibility for both the content (what is to be accomplished) and the process (how it is to be accomplished). Leaders must receive sponsor sign-off before moving from one step to the next in the problem-solving model. The team leader may run the meeting or may wish to participate as a group member and therefore ask the group facilitator to conduct the meeting.

Facilitator

The facilitator is normally an individual from outside the team (another unit or department). This individual focuses on process. The facilitator helps the team leader to prepare for meetings and to assist in the implementation of the problem-solving model. He or she is the meeting chauffeur, a servant to the group. The facilitator is neutral and nonevaluative.

Recorder

The recorder is selected from the team. This responsibility is normally rotated from one team member to another. It is this individual's responsibility to record the "group memory," and, like the facilitator, the recorder is a neutral, nonevaluative servant of the group. Recorders write down in full view of the team (perhaps on a large chart) enough detail so that ideas can be preserved and reviewed at a later time.

Team Members

Individual members of a TQ team are normally selected due to a particular area of responsibility or expertise. They contribute both information and ideas. Effective team members are supportive and constructive because one important aspect of TQ focuses on the strength of the team as a group.

The possibility of you as an individual engineering student being part of team-related activities is high, so you should learn as much as possible about team interaction. Working together as a team on a project provides a wide array of advantages; however, the team environment is complex. If the team mix—that is, personality, chemistry, background, and knowledge—is proper, a great deal of enthusiasm, esprit de corps, and "fun," not to mention progress, can be derived from the team experience. On the other hand, if you have no knowledge of team dynamics, or if you have no training in facilitative leadership, you may not be an effective member. As a unit you will more than likely waste time on struggles for control and endless discussion, with little or no meaningful progress.

The more you can learn about what to expect or what to anticipate as you are placed on a team, the better equipped you will be to handle difficult situations. You must be able to recognize and avoid disruptive behavior and learn to work through those situations that cannot be avoided.

Team Dynamics

Most college freshmen have worked with other students in various team environments on specific projects. This may have occured in secondary school, sports, church, or during some other social activities. Typically when you are placed together as a college team, it is likely that the members will consist of a nonhomogeneous mixture of personalities and cultural backgrounds. For example, a design team may consist of someone from your own hometown, another individual from a different part of the state, an international student, or perhaps an adult student. This is both good and bad. It is good if you understand that the best solution to problems will be found from within a mixture of opinions, backgrounds, experiences, ideas, and cultural diversity. It is bad when you do not understand and accept the fact that input from other members may contrast with your own ideas, thought processes, and beliefs. You must learn to appreciate the talent and value of all team members and to open your mind to possible approaches that do not parallel your background and experience.

The moment you are placed within a team environment a series of events take place that are predictable. The following is adapted from work by Bruce W. Tuckman,[8] who suggests that team growth develops in the following four unique stages.

Stage 1: Forming.   During this stage members are tentative, exploring the limits of acceptable group behavior. The forming period includes excitement and anticipation, a feeling of pride that they have been selected for this assignment. However, it also includes serious suspicion, fear, and anxiety about the project. Teams begin the process of defining the task at hand and decide how it might be accomplished. They begin an informal search for acceptable group behavior and ponder how to deal with group problems. It is likely that discussion will be tentative, suggesting abstract thoughts, concepts, and issues. Because members are cautious and tentative, and individuals are trying to size up other members within the team, the initial or forming stage does not result in much progress. If this happens, it is perfectly normal.

Stage 2: Storming.   Typically the team transitions from little noticeable progress to downright trouble. Storming is the most difficult stage. A swimming analogy provides a good example. Team members believe they have been tossed in the water and think they are going to drown. Members thrash around because they see the task as formidable. They are not at all sure how deep the water is. They slowly begin to form a global vision of the task as a team but still rely on their own personal and professional experiences, resisting the many advantages of collaborative team work. The storming stage includes a resistance to the introduction of new tools that are available because these new approaches suggest methods that are at odds with your own background and experiences. There is sharp disagreement among members with frequent hostility. During this stage members tend to be defensive and competitive. It is very unlikely that much time or energy is being devoted to problem solving. This stage is typical and perhaps even necessary for team dynamics to evolve. Not much work is accomplished, but, once again, this is normal.

Stage 3: Norming.   During this stage events start to crystallize. Members accept the "team ground rules." They begin to establish their roles as team members and slowly to begin the process of helping each other and working together. Norming is the stage of accepting team membership, developing a belief that the project is possible, and expressing criticism constructively. Members become more confident in one another and develop a general sense of team

---

[8]Bruce W. Tuckman, "Development Sequence in Small Groups," *Psychological Bulletin* Vol. No. 15, 1965.

cohesion, a common spirit and goal. As the team members work out differences, they spend more energy on problem resolution. They actually begin to see progress.

Stage 4: Performing.   This last stage of team dynamics suggests that the team has finally resolved and established its relationships. The team arrives at the performing step by real problem solving. Members have discovered and accepted each other's strengths and shortcomings. The team is a cohesive unit capable of getting a lot of work done.

These stages are normal, and a member who understands that all new teams must go through these steps in some fashion will be better equipped to deal with them.

Teams that have not received TQ training or teams that have not been involved in previous TQ projects are not as likely to be successful. Without prior training and planning a great deal of time and energy will be wasted. With this thought in mind, we have provided the following material to help you get started in the team environment.

Team Formation

When you are initially placed on a team, it can be difficult and sometimes stressful. You may not know any of the team members, and, lacking very specific instructions, it is not easy to know what to do or how to get started. To help you through this situation, let us review again the "forming" stage. During this stage members are cautious; they begin the process of defining the task at hand and discuss how it might be accomplished. It is unlikely that meaningful progress will be made during this stage, which is perfectly normal. Remember, the team is a powerful unit when properly organized, motivated, empowered, and managed. Individual members who are collectively focused on a well-defined outcome can accomplish amazing results.

Team activity is normally divided into two components: content and process. Content is the "what" that is to be accomplished and process is the "how" it will get done. Team members are either working on a specific assignment as individual members or they are collected together in a team meeting.

Team Meetings

As a professional you will spend a large percentage of your time in meetings. Some of these meetings will be well organized and some will not. It has been estimated that throughout the United States in a single day there are 10 million meetings. As a new engineer you may spend 25 percent of your time in meetings, and as you become a part of middle and top management, you can spend as much as 50 percent of your time in meetings. It is safe to say meetings consume tremendous amount of time, money, and energy, resources that should be carefully managed.

Let us return to the TQ team and use the organizational structure that we have developed to discuss team meetings. First, a successful team must have a team leader. To most people the term "leader" typically means the one who is "in charge." Since we do not especially want the leader to be in charge, a different title for this individual is used in the TQ domain: facilitative team leader. This individual is not in charge in the sense that he or she is the czar or dictator. The facilitator team leader is just as the name implies, he or she is to make sure that the meeting is well organized. This person has a specific set of responsibilities just as the recorder and team members have specific duties. Every member of the team should acquire the training and skill to be an effective facilitative team leader, particularly if the member is an aspiring engineering student.

The first important function of the facilitative team leader is to plan the meeting. There are different types of meetings that can be held. Three possible varieties are:

> *Informational meeting.* The purpose of this type of meeting is to distribute or gather information.
>
> *Action-oriented meeting.* The purpose of this type of meeting is to take action by the group; that is, to plan, problem solve, or make a decision.
>
> *Combination meeting.* Many meetings are a combination of these two; however, it is important that team members know before the meeting if a decision will be made or an action item will take place.

Once the facilitative team leader has decided on the type of meeting, the next order of business is to plan the desired outcome—that is, to decide what will happen as a result of the meeting. It is important to establish a clear and concise statement of the desired outcome. Having defined desired outcomes provides direction for developing the meeting agenda. At the end of the meeting the team members should be able to determine if they have accomplished the desired outcome. If the outcomes of a meeting are not clear, it is more than likely they will not be accomplished.

An effective facilitative team leader will prepare an agenda and distribute it to the team members with sufficient lead time for review and preparation by members. The facilitator who is assigned to assist the leader will arrange meeting space, as well as arrange for any special requirements such as flip charts, overhead projectors, and so on.

The more effective and complete the preparation is, the more productive the meeting will be. In general there is a one-to-one ratio between preparation and meeting time. If you are planning a 2-hour meeting, it will require 2 hours of preparation.

A facilitative team leader may start a meeting as follows:

> I have convened this meeting to provide new information, gather some information from previous assignments of the membership, and to decide on a unique course of action from several possible actions listed on the flip chart at the front of the room (combination meeting). The desired outcome will be as follows . . . [whatever it is]. I'll be facilitating the meeting and I'd like to ask John Doe to serve as team recorder [that function is normally assigned and then rotated]. We'll make all decisions by consensus, and I'll be the fallback if we don't reach consensus in the time allowed.

It is extremely important to establish the decision-making process. Many teams do not carry this out, and it is a big mistake. There are different ways that decisions can be made:

1. Decide and announce—if a team leader in an organization also happens to be the "boss" or has been delegated considerable authority, then this form of decision making may be appropriate.

2. Gather information from individuals or team members and then the leader decides.
3. Gather information from individuals or team members and then the team decides by majority vote.
4. Make a decision by consensus.
5. Delegate the decision to membership with criteria.

We will focus on the fourth method: decision making by consensus. This concept is not clearly understood by most people. Consensus does not mean everyone agrees, nor is it a vote whereby majority rules. Consensus means everyone agrees to support the group decision.

As stated by William G. Ouchi, author of *Theory Z,* consensus has been reached when all team members can agree on a single solution or decision and each member can say:

■ I believe that each of you have listened to and understand my point of view.
■ I believe that I understand your point of view.
■ I may or may not prefer this decision, but I will support it because it was reached openly and fairly.

In other words, a consensus decision is one with which each team member can live and actively support—it is not necessarily each member's number one choice.

This concept leads us to another critical issue for successful team meetings: the idea of "ground rules." Any well-functioning team must establish its own set of operational ground rules. The following is a set of typical ground rules for conducting a meeting:

1. All decisions will be by consensus.
2. Sensitive issues are confidential.
3. Listen to what others have to say.
4. Be prepared for the meeting.
5. Be on time.
6. Contribute actively.
7. Improve team relationships whenever possible.
8. Support the TQ process.
9. Keep records of your work.
10. Be flexible.
11. Do not dominate the meeting.
12. A minimum of four team members must be present to conduct a meeting.

It is important to understand the other responsibilities an effective facilitative team leader has beyond planning. Two additional key elements in any successful meeting are conducting the meeting itself and follow-up. Let us look at these two items separately.

Conducting the Meeting. The facilitative team leader can run the meeting or can elect to be a participant during the meeting and invite the team "facilitator" to run the meeting. Recall that a TQ team has an external facilitator. Again, this is a person from outside the organization who is designated to assist the facilitative team leader with process.

As the meeting begins keep in mind that all teams will have their share of problems. Remember, the second stage of team development was "storming." Recall that in the beginning a team's normal transition is from little noticeable progress to trouble. Storming is a difficult stage. The team facilitator is trained to help, but the individual team members are the ones that must make the team work. It may be necessary at different times, as a team, to work through a particular problem.

For example, providing feedback to other team members is necessary on occasion, but again this step must be done with care.

It is likely that the team will experience some or all the following common problems, and there are methods and techniques developed to assist in their resolution. Common team or group problems are:

- Stagnation
- Overbearing or dominating members
- Reluctant participants
- Unquestioned acceptance of opinion as fact
- Rush to accomplishments
- Feuding team members

All individuals on the team are different, and sometimes the most important thing to do is to listen carefully to what others have to say. Only then can we begin to understand people and effectively to deal with individuals who are not just like us.

Meeting Follow-up

Once the meeting is over there are a number of important activities that remain to be accomplished. It is extremely likely that during the course of the meeting there were a number of action items that need attention and follow-up. The minutes of the meeting must be distributed, and the team members may need reminders of things they were asked to do. The facilitative team leader may need to receive information or to get approval from the sponsor.

1.8.3   Personal Style Inventory

We have indicated throughout this chapter, in fact, have stressed the idea that all individuals are not the same. Various methods are available to explore these individual differences, and one very common approach is referred to as a personal style inventory assessment. A personal style inventory assessment allows each of us as individuals to understand our own personalities and how they differ from others in the team. The most beneficial information derived from personal inventories is a clear understanding that everyone does not think and act the same. Good, creative, hard working, ethical people think, act, and react differently.

In 1986 Robert W. Russell developed a personal inventory test, which is a set of questions that can be answered to determine one's personality category. For example, do you like things to move with a "fast pace" or more "slowly and deliberately"? Do you "tend to be impatient" or do you display "a great deal of patience"? Your response to a series of questions of this type tend to categorize you as either:

Task oriented or people oriented
Reflective or assertive

Individuals who are task oriented and reflective tend to be analyzers. Those who are people oriented and assertive are enthusiasts.

Glenn M. Parker, in his 1990 book *Team Players and Teamwork,* categorizes four team player styles:

- Contributor

  They tend to provide all relevant knowledge, skills, and data. A contributor sees the team as a group of subject matter experts.

  A contributor checklist: dependable, responsible, organized, efficient, logical, clear, relevant, pragmatic, systematic, and proficient.

- Collaborator

  They seek an overall vision. The project goals are paramount. They are willing to work outside a defined role to get the work done.

  A collaborator checklist: cooperative, flexible, confident, forward-looking, conceptual, accommodating, generous, open, visionary, and imaginative.

- Communicator

  They tend to be a process-oriented team member. They can be effective listeners and facilitators or function as participants focused on conflict resolution, consensus building, feedback, and the building of a relaxed climate.

  A communicator checklist: supportive, encouraging, relaxed, tactful, helpful, friendly, patient, informal, considerate, and spontaneous.

- Challenger

  They openly question process, vision, goals, and other team members.

  A challenger checklist: candid, ethical, questioning, honest, truthful, outspoken, principled, adventurous, aboveboard, and brave.

MBTI (Myers Briggs Type Indicator)

Katharine Briggs and Isabel Briggs Myers conceived the idea for a personality indicator in the early 1940s, and it has evolved into a reasonably sophisticated instrument. The actual MBTI consists of 200-plus questions that require considerable time to complete. Personality questionnaires such as the MBTI have been used for many years. They are a popular method of assessment.

R. Craig Hogan and David W. Champagne have derived a shorter version of the MBTI instrument that can be completed in much less time. This particular inventory is self-administering and self-scoring. It takes about 10 minutes to fill out and 5 minutes to score. Items are arranged in pairs (a and b), and each member of the pair represents a preference you may or may not hold. Individuals are asked to rate their preference for each item by giving it a score of 0 to 5 (0 meaning you *really* feel negative about it or strongly about the other member of the pair, 5 meaning you *strongly* prefer it or do not prefer the other member of the pair). The scores for (a) and (b) *must add up to* 5 (0 and 5, 1 and

59

Problem Solving and the Team

4, 2 and 3, etc.).

A few sample pairs are listed here:[9]

I prefer:

1a. _____ Making decisions after finding out what others think.
1b. _____ Making decisions without consulting others.
2a. _____ Being called imaginative or intuitive.
2b. _____ Being called factual and accurate.
3a. _____ Making decisions about people in organizations based on available data and systematic analysis of situations.
3b. _____ Making decisions about people in organizations based on empathy, feelings, and understanding of their needs and values.

As a team member each of you brings a range of experiences and values to the team environment. In so doing you can promote either congenial, comfortable, productive discussion or frustrating, conflicting, unproductive argument that reflects your needs and prejudices rather than the real issues of the project.

When people who interact with each other understand their own values and unique backgrounds and why these affect their thinking and interaction, they likely will be more constructive about the ideas and suggestions they advance, seeing them as ideas they value rather than as absolutes. They also will be more able to accept the ideas or actions of others that differ from their own.

The purpose of a personal style inventory is to enable individuals to identify their personality types so that they can learn to understand better the influence of personality style on their thoughts and actions and on the thoughts and actions of those with whom they interact. An inventory assessment also will help to identify strengths and weaknesses in their own styles.

The MBTI personal style inventory classifies four pairs of personality traits:

Introversion—extroversion
Intuition—sensing
Feeling—thinking
Perceiving—judging

Every individual exhibits all eight in thought and action, but for each person one personality trait of each pair is used more often, is more comfortable, and has given rise to a greater number of beliefs and values than has the other member of the pair. The trait, consequently, characterizes the person's outlook and thought processes. As a result, a person's type of preference for four of these traits (one from each of the four pairs) can be determined, and predictions about values, beliefs, and behavior can be made based on the results.

When participants understand how profoundly these personality traits affect their values, choices, assumptions, beliefs, decisions, thoughts, and behavior and

---

[9]The complete survey can be found in R. C. Hogan and D. W. Champagne, "Personal Style Inventory," *1980 Annual Handbook for Group Facilitators,* San Diego, CA: University Associates, 1980, pp. 89–99. Although the material is copyrighted, it may be freely reproduced for educational/training/research activities.

those of their spouses, colleagues, superiors, subordinates, students, and instructors, then they can begin to realize that the statements and actions that they and those around them live by are the result of different views of the world, not right or wrong thinking. Our own point of view and those of others can be more easily understood and accepted.

Personality Traits

These personality traits are present to some degree in all people. It is the extremes that are described here. Hogan and Champagne describe the pairs of traits as follows:

Introversion—Extroversion. People who are more introverted than extroverted tend to make decisions somewhat independently of constraints and prodding from the situation, culture, people, or things around them. They are quiet, diligent at working alone, and socially reserved. They may dislike being interrupted while working and may tend to forget names and faces.

Extroverts are attuned to the culture, people, and things around them, endeavoring to make decisions congruent with demands and expectations. The extrovert is outgoing, socially adept, and interested in variety and in working with people. The extrovert may become impatient with long, slow tasks and does not mind being interrupted by people.

Intuition—Sensing. Intuitive people prefer possibilities, theories, gestalts, the overall, invention, and the new; they become bored with nitty-gritty details, the concrete and actual data, and facts unrelated to concepts. Intuitive people think and discuss in spontaneous leaps of intuition that may leave out or neglect details. Problem solving comes easily for these individuals although there may be a tendency to make errors of fact.

The sensing types prefer the concrete, real, factual, structured, tangible here and now, becoming impatient with theory and the abstract, mistrusting intuition. The sensing type thinks in careful, detail-by-detail accuracy, remembering real facts, making few errors of fact but possibly missing a conception of the overall.

Feeling—Thinking. The feelers make judgments about life, people, occurrences, and things based on empathy, warmth, and personal values. As a consequence, feelers are more interested in people and feelings than in impersonal logic, analysis, and things, and in conciliation with harmony more than in being on top or achieving impersonal goals. The feelers get along well with people in general.

The thinkers make judgments about life, people, occurrences, and things based on logic, analysis, and evidence, avoiding the irrationality of making decisions based on feelings and values. As a result, the thinkers are more interested in logic, analysis, and verifiable conclusions than in empathy, values, and personal warmth. The thinkers may "step" on others' feelings and needs without realizing it, neglecting to take into consideration the values of others.

Perceiving—Judging. The perceivers are gatherers, always wanting to know more before deciding, holding off decisions and judgments. As a consequence, perceiving people are open, flexible, adaptive, nonjudgmental, able to see and appreciate all sides of issues, and always welcoming new perspectives and new information about issues. However, perceivers also are difficult to pin down and may be indecisive and noncommittal, becoming involved in many tasks that do not reach closure so that they may become frustrated at times. Even when they finish tasks, perceivers will tend to look back at them and wonder whether they are satisfactory or could have been done another way. The perceiver wishes to roll with life rather than change it.

The judges are decisive, firm, and sure, setting goals and sticking to them. The judges want to close books, to make decisions, and to get on to the next project. When a project does not yet have closure, judges will leave it behind and go on to new tasks and not look back.

The Team Mix

The purpose of this overview is to point out that each team member has different strengths. In the beginning you will perceive these differences as time-consuming irritations. Imagine someone challenging something you have stated or presented—the nerve of that individual!

The storming stage of team evolution can be less disastrous if we understand that all of us have different styles. The categorization of these styles, however, is limited and somewhat arbitrary. Individuals are far too complex to be labeled. We do not wish to stereotype people too easily. There is no right or wrong style—all have strengths and weaknesses that make them no better or worse than each other. Styles are often situational and may change depending on mood, environment, and state of mind. To be successful, teams need a mix which provides a strength for task accomplishment and group process.

This brief introduction to TQ demonstrates that this process is basically one of common sense. The key to success of an organization is its people and how they interact to accomplish the desired goals. As a practicing engineer you will be at the heart of the organization performing one or more of the functions of design, development, production, operations, construction, sales, management, and research. Your ability to work with people, apply problem-solving techniques, and work within specified time and budget constraints will determine your success. Total quality is a philosophy that can help you to maximize your success. Let us now devote our attention to the education of the engineer and the elements of professionalism and ethics in engineering practice.

1.9　Education of the Engineer

The amount of information coming from the academic and business world is increasing exponentially, and at the current rate it will double in less than 20 years. More than any other group, engineers are using this knowledge to shape civilization. To keep pace with a changing world, engineers must be educated to solve problems that are as yet unheard of. A large share of the responsibility for this mammoth education task falls on the engineering colleges and universities. But the completion of an engineering program is only the first

Figure 1.32

With the World Wide Web, information from all over the world is available in your home or at work at a computer. Many universities and businesses are providing formal classwork using the WWW.

step toward a lifetime of education. The engineer, with the assistance of the employer and the university, must continue to study. See Fig. 1.32.

Logically, then, an engineering education should provide a broad base in scientific and engineering principles, some study in the humanities and social sciences, and specialized studies in a chosen engineering curriculum. But specific questions concerning engineering education still arise. We will deal here with the questions that are frequently asked by students. What are the desirable characteristics for success in an engineering program? What knowledge and skills should be acquired in college? What is meant by continuing education with respect to an engineering career?

1.9.1   Desirable Characteristics

Years of experience have enabled engineering educators to analyze the performance of students in relation to the abilities and desires they possess when entering college. The most important characteristics for an engineering student can be summarized as follows:

1.   A strong interest in and ability to work with mathematics and science taken in high school.

Table 1.5 Participating Bodies in the Accreditation Activity

Organization	Abbreviation
American Academy of Environmental Engineers	AAEE
American Congress on Surveying and Mapping	ACSM
American Institute of Aeronautics and Astronautics, Inc.	AIAA
American Institute of Chemical Engineers	AICHE
American Nuclear Society	ANS
American Society for Engineers Education	ASEE
American Society of Agricultural Engineers	ASAE
American Society of Civil Engineers	ASCE
American Society of Heating, Refrigerating, and Air-Conditioning Engineers, Inc.	ASHRAE
The American Society of Mechanical Engineers	ASME
The Institute of Electrical and Electronics Engineers, Inc.	IEEE
Institute of Industrial Engineers, Inc.	IIE
International Society for Measurement and Control	ISA
The Minerals, Metals & Materials Society	TMS
National Council of Examiners for Engineering and Surveying	NCEES
National Institute of Ceramic Engineers	NICE
National Society of Professional Engineers	NSPE
Society for Mining, Metallurgy, and Exploration, Inc.	SME-AIME
Society of Automotive Engineers	SAE
Society of Manufacturing Engineers	SME
Society of Naval Architects and Marine Engineers	SNAME
Society of Petroleum Engineers	SPE

Source: Compiled from ABET Annual Report, 1994.

2. An ability to think through a problem in a logical manner.
3. A knack for organizing and carrying through to conclusion the solution to a problem.
4. An unusual curiosity about how and why things work.

Although such attributes are desirable, having them is no guarantee of success in an engineering program. Simply a strong desire for the job has made successful engineers of some individuals who did not possess any of these characteristics; and, conversely, many who possessed them did not complete an engineering degree. Moreover, an engineering education is not easy, but it can offer a rewarding career to anyone who accepts the challenge.

1.9.2 Knowledge and Skills Required

As indicated previously, over 340 colleges and universities offer programs in engineering that are accredited by ABET or CEAB. These boards have as their purpose the quality control of engineering and technology programs offered in the United States and Canada. The basis of the boards is the engineering profession, which is represented through the participating professional groups. A listing of the participating bodies is given in Table 1.5.

The quality control of engineering programs is effected through the accreditation process. The engineering profession, through ABET and CEAB, has

Figure 1.33

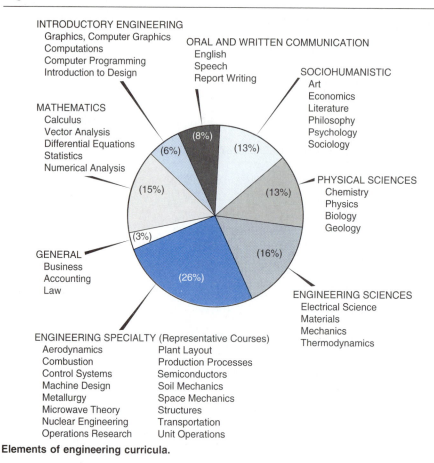

INTRODUCTORY ENGINEERING
  Graphics, Computer Graphics
  Computations
  Computer Programming
  Introduction to Design

ORAL AND WRITTEN COMMUNICATION
  English
  Speech
  Report Writing

SOCIOHUMANISTIC
  Art
  Economics
  Literature
  Philosophy
  Psychology
  Sociology

MATHEMATICS
  Calculus
  Vector Analysis
  Differential Equations
  Statistics
  Numerical Analysis

PHYSICAL SCIENCES
  Chemistry
  Physics
  Biology
  Geology

GENERAL
  Business
  Accounting
  Law

ENGINEERING SCIENCES
  Electrical Science
  Materials
  Mechanics
  Thermodynamics

(8%) (13%) (13%) (6%) (15%) (3%) (16%) (26%)

ENGINEERING SPECIALTY (Representative Courses)
  Aerodynamics   Plant Layout
  Combustion   Production Processes
  Control Systems   Semiconductors
  Machine Design   Soil Mechanics
  Metallurgy   Space Mechanics
  Microwave Theory   Structures
  Nuclear Engineering   Transportation
  Operations Research   Unit Operations

Elements of engineering curricula.

developed standards and criteria for the education of engineers entering the profession. Through visitations, evaluations, and reports the written criteria and standards are compared with the engineering curricula at a university. Each program, if operating according to the standards and criteria, may receive up to 6 years of accreditation. If some discrepancies appear, accreditations may be granted for a shorter time period or may not be granted at all until appropriate improvements are made.

It is safe to say that for any given engineering discipline, no two schools will have identical offerings. However, close scrutiny will show a framework within which most courses can be placed, with differences occurring only in textbooks used, topics emphasized, and sequences followed. Figure 1.33 depicts this framework and some of the courses that fall within each of the areas. The approximate percentage of time spent on each course grouping is indicated.

The sociohumanistic block is a small portion of most engineering curricula, but it is important because it helps the engineering student to understand and develop an appreciation for the potential impact of engineering to undertakings on the environment and general society. When the location of a nuclear power plant is being considered, the engineers involved in this decision must respect the concerns and feelings of all individuals who might be affected by the location. Discussions of the interaction between engineers and the general public take place in few engineering courses; sociohumanistic courses thus are needed to furnish engineering students with an insight into the needs and aspirations of society.

Chemistry and physics are almost universally required in engineering. They are fundamental to the study of engineering science. The mathematics normally required for college chemistry and physics is more advanced than that for the corresponding high school courses. Higher-level chemistry and physics also may be required, depending on departmental structure. Finally, other physical science courses may be required in some programs or taken as electives.

An engineer cannot be successful without the ability to communicate ideas and the results of work efforts. The research engineer writes reports and orally presents ideas to management. The production engineer must be able to converse with craftspersons in understandable terms. And all engineers have dealings with the public and must be able to communicate on a nontechnical level.

Engineers have been accused of not becoming involved in public affairs. The reason often given for this noninvolvement is that they are not trained sufficiently in oral and written communications. However, the equivalent of one-third of one year is spent on formal courses in these subjects, and additional time is spent in design presentations, written laboratory reports, and the like. A conscious effort by student engineers must be made to improve their abilities in oral and written communication to overcome this nonactivist label.

Mathematics is the most powerful tool that the engineer uses to solve problems. The amount of time spent in this area is indicative of its importance. Courses in calculus, vector analysis, and differential equations are common to all degree programs. Statistics, numerical analysis, and other mathematics courses support some engineering specialty areas. Students desiring an advanced degree may want to take mathematics courses beyond the baccalaureate-level requirements.

In the early stages of an engineering education introductory courses in graphical communication, computational techniques, design, and computer programming are taken. Engineering schools vary somewhat in their emphasis on these areas, but the general intent is to develop skills in the application of theory to practical problem solving and familiarity with engineering terminology. Design is presented from a conceptual point of view to aid the student in creative thinking. Graphics develops the visualization capability and assists the student in transferring mental thoughts into well-defined concepts on paper. The tremendous potential of the computer to assist the engineer has led to the requirement of computer programming in almost all curricula. Use of the computer to perform many tedious calculations has increased the efficiency of the engineer and has allowed more time for creative thinking. Computer graphics is becoming a part of the engineering curricula. Its ability to enhance the visualization of geometry and to depict engineering quantities graphically has increased productivity in the design process.

With a strong background in mathematics and physical sciences, you can begin study to engineering sciences, courses that are fundamental to all engineering specialties. Electrical science includes study of charges, fields, circuits, and electronics. Materials science courses involve study of the properties and

Figure 1.34

New materials, such as the metals in a high-speed aircraft turbofan engine, are developed by materials engineers.

chemical compositions of metallic and nonmetallic substances (see Fig. 1.34). Mechanics includes study of statics, dynamics, fluids, and mechanics of materials. Thermodynamics is the science of heat and is the basis for study of all types of energy and energy transfer. A sound understanding of the engineering sciences is most important for anyone interested in pursuing postgraduate work and research.

Figure 1.33 shows only a few examples of the many specialized engineering courses given. Scanning course descriptions in a college general bulletin or catalog will provide a more detailed insight into the specialized courses required in the various engineering disciplines.

Most curricula allow a student flexibility in selecting a few courses in areas that were not previously mentioned. For example, a student interested in management may take some courses in business and accounting. Another may desire some background in law or medicine, with the intent of entering a professional school in one of these areas upon graduation from engineering.

1.10 Professionalism and Ethics

Engineering is a learned vocation, demanding an individual with high standards of ethics and sound moral character. When making judgments that may create controversy and affect many people, the engineer must keep foremost in mind a dedication to the betterment of humanity.

The engineering profession has attempted for many years to become unified. However, technical societies that represent individual engineering disciplines have grown strong and tend to keep the various engineering disciplines separated. This contrasts with more unified professions such as law, medicine, and theology.

1.10.1   Professionalism

Professionalism is a way of life. A professional person is one who engages in an activity that requires a specialized and comprehensive education and is motivated by a strong desire to serve humanity. A professional thinks and acts in a manner that brings favor upon the individual and the entire profession. Developing a professional frame of mind begins with your engineering education.

The professional engineer can be said to have the following:

1. Specialized knowledge and skills used for the benefit of humanity.
2. Honesty and impartiality in engineering service.
3. Constant interest in improving the profession.
4. Support of professional and technical societies that represent the professional engineer.

It is clear that these characteristics include not only technical competence but also a positive attitude toward life that is continually reinforced by educational accomplishments and professional service.

A primary reason for the rapid development in science and engineering is the work of technical societies. The fundamental service provided by a society is the sharing of ideas, which means that technical specialists can publicize their efforts and assist others in promoting excellence in the profession. When information is distributed to other society members, new ideas evolve and duplicated efforts are minimized. The societies conduct meetings on international, national, and local bases. Students of engineering will find a technical society in their specialty that may operate as a branch of the regular society or as a student chapter on campus. The student organization is an important link with professional workers, providing motivation and the opportunity to make acquaintances that will help students to formulate career objectives.

Table 1.5 is a partial listing of the numerous engineering societies that support the engineering disciplines and functions. These technical societies are linked because of their support of the accreditation process. Over 60 other societies exist for the purpose of supporting the professional status of engineers. Among these are the Society of Women Engineers (SWE), the National Society of Black Engineers (NSBE), the Acoustical Society of America (ASA), the Society of Plastics Engineers (SPE), and the American Society for Quality Control (ASQC).

Many technical societies are quite influential in the engineering profession. To unify the profession in the manner of such other professions as law, medicine, and theology would require the cooperation of all the societies. However, the individual societies have more than satisfied the professional needs for many engineers, so no pressing desire to unify is apparent. Nonetheless, to

preserve the advantages of the technical societies while unifying the entire profession remains a long-range goal of engineers today.

1.10.2 Professional Registration

The power to license engineers rests with each of the 50 states. Since the first
registration law in Wyoming in 1907 all states have developed legislation specifying requirements for engineering practice. The purpose of registration laws is
to protect the public. Just as one would expect a physician to provide competent medical service, an engineer can be expected to provide competent technical service. However, the laws of registration for engineers are quite different
from those for lawyers or physicians. An engineer does not have to be registered
to practice engineering. Legally, only the chief engineer of a firm needs to be
registered for that firm to perform engineering services. Individuals testifying
as expert engineering witnesses in court and those offering engineering consulting services need to be registered. In some instances the practice of engineering is allowed as long as the individual does not advertise as an engineer.

The legal process for becoming a licensed professional engineer consists
of four parts, two of which entail examinations. The parts include:

1. An engineering degree from an acceptable institution as defined by the
state board for registration. Graduation from an ABET-accredited institution
satisfies the degree requirement automatically.
2. Successful completion of the Fundamentals of Engineering Examination
(FE) entitles one to the title "engineer-in-training" (EIT). This 8-hour examination may be taken during the last term of an undergraduate program that
is ABET-accredited. The first half of the exam covers fundamentals in the
areas of mathematics, chemistry, physics, engineering mechanics, electrical science, thermal science, economics, and ethics. The second half is oriented to
each discipline such as mechanical. The passing grade is determined by the
state board.
3. Completion of 4 years of engineering practice as an EIT.
4. Successful completion of the Principles and Practice Examination completes the licensing process. This also is an 8-hour examination covering problems normally encountered in the area of specialty such as mechanical or chemical engineering.

It should be noted that once the license is received, it is permanent although
there is an annual renewal fee. In addition, the trend is toward specific requirements in continuing education each year in order to maintain the license.
Licensed engineers in some states may attend professional meetings in their
specialty, take classes, and write professional papers or books to accumulate
sufficient professional development activities beyond their job responsibilities
to maintain their licenses. This trend is a reflection of the rapidly changing
technology and the need for engineers to remain current in their area.

Registration does have many advantages. Most public employment positions, all expert witness roles in court cases, and some high-level company
positions require the professional engineer's license. However, less than one-
half the eligible candidates are currently registered. You should give serious
consideration to becoming registered as soon as you qualify. Satisfying the

requirements for registration can be started even before graduation from an ABET-accredited curriculum.

1.10.3 Professional Ethics

Ethics is the guide to personal conduct of a professional. Most technical societies have a written code of ethics for their members. Because of this, some variations exist; but a general view of ethics for engineers is provided here for two of the technical societies. Figure 1.35 is a code endorsed by the ABET. Appendix D gives the most widely endorsed code of ethics, that of the National Society of Professional Engineers (NSPE). As you read both codes, note the many similarities. Figure 1.36 is the "Engineer's Creed" as published by the NSPE.

1.11 Challenges of the Future

The world continues to undergo rapid and sometimes tumultuous change. As a practicing engineer, you will occupy center stage in many of these changes in the near future and will become even more involved in the more distant future. The huge tasks of providing energy, maintaining a supply of water, ensuring a competitive edge in the world marketplace, rebuilding our infrastructure, and preserving our environment will challenge the technical community beyond anyone's imagination.

Engineers of today have nearly instantaneous access to a wealth of information from technical, economic, social, and political sources. A key to the success of engineers in the future will be the ability to study and absorb the appropriate information in the time allotted for producing a design or solution to a problem. A degree in engineering is only the beginning of a lifelong period of study in order to remain informed and competent in the field.

Engineers of tomorrow will have even greater access to information and will use increasingly powerful computer systems to digest this information. They will work with colleagues around the world solving problems and creating new products. They will assume greater roles in making decisions that affect the use of energy, water, and other natural resources.

1.11.1 Energy

In order to develop technologically, nations of the world require vast amounts of energy. With a finite supply of our greatest energy source, fossil fuels, alternate supplies must be developed and existing sources must be controlled with a worldwide usage plan. A key factor in the design of products must be minimum use of energy.

We do not mean to imply that our fossil fuel resources will be gone in a short time. However, as demand increases and supplies become scarcer, the cost of obtaining the energy increases and places additional burdens on already financially strapped regions and individuals. Engineers with great vision are needed to develop alternative sources of energy from the sun, wind, and ocean and to improve the efficiency of existing energy consumption devices.

Figure 1.35

Accreditation Board for Engineering and Technology*

CODE OF ETHICS OF ENGINEERS

THE FUNDAMENTAL PRINCIPLES

Engineers uphold and advance the integrity, honor and dignity of the engineering profession by:

I. using their knowledge and skill for the enhancement of human welfare;

II. being honest and impartial, and serving with fidelity the public, their employers and clients;

III. striving to increase the competence and prestige of the engineering profession; and

IV. supporting the professional and technical societies of their disciplines.

THE FUNDAMENTAL CANONS

1. Engineers shall hold paramount the safety, health and welfare of the public in the performance of their professional duties.

2. Engineers shall perform services only in the areas of their competence.

3. Engineers shall issue public statements only in an objective and truthful manner.

4. Engineers shall act in professional matters for each employer or client as faithful agents or trustees, and shall avoid conflicts of interest.

5. Engineers shall build their professional reputation on the merit of their services and shall not compete unfairly with others.

6. Engineers shall act in such a manner as to uphold and enhance the honor, integrity and dignity of the profession.

7. Engineers shall continue their professional development throughout their careers and shall provide opportunities for the professional development of those engineers under their supervision.

345 East 47th Street New York, NY 10017

*Formerly Engineers' Council for Professional Development. (Approved by the ECPD Board of Directors, October 5, 1977)

AB-54 2/85

Code of Ethics for Engineers. (*Accreditation Board for Engineering and Technology*)

Figure 1.36

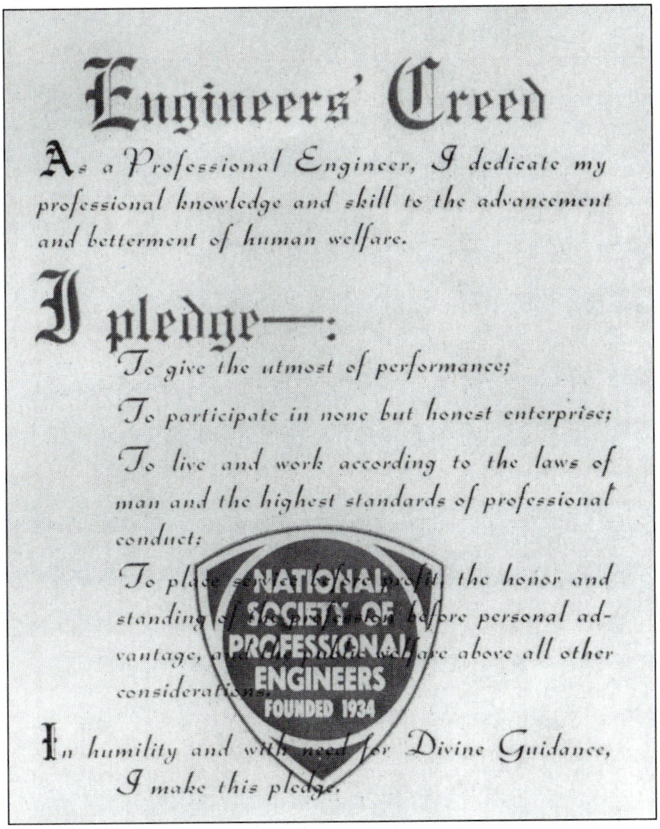

71

*Challenges of
the Future*

Engineer's Creed. (*National Society of Professional Engineers*).

Along with the production and consumption of energy come the secondary problems of pollution. Such pollutants as smog and acid rain, carbon monoxide, and radiation must receive attention in order to maintain the balance of nature.

1.11.2 Water

The basic water cycle—from evaporation to cloud formation, then to rain, runoff, and evaporation again—is taken for granted by most people. However, if the rain or the runoff is polluted, then the cycle is interrupted and our water supply becomes a crucial problem. In addition, some highly populated areas have a limited water supply and must rely on water distribution systems from other areas of the country. Many formerly undeveloped agricultural regions are now productive because of irrigation systems. However, the irrigation systems deplete the underground streams of water that are needed downstream.

These problems must be solved in order for life to continue to exist as we know it. Because of the regional water distribution patterns, the federal government must be a part of the decision-making process for water distribution.

Figure 1.37

Coal-fired generating stations produce a large share of our electricity.

One of the concerns that must be eased is the amount of time required to bring a water distribution plan into effect. Government agencies and the private sector are strapped by regulations that cause delays in planning and construction of several years. Greater cooperation and a better informed public are goals that public works engineers must strive to achieve.

Developing nations around the world need an additional water supplies because of increasing populations. Many of these nations do not have the necessary freshwater and must rely on desalination, a costly process. The continued need for water is a concern for leaders of the world, and engineers will be asked to create additional sources of this life-sustaining resource.

1.11.3 A Competitive Edge in the World Marketplace

We have all purchased or used products that were manufactured outside the country. Many of these products incorporate technology that was developed in the United States. In order to maintain our strong industrial base, we must develop practices and processes that enable us to compete, not just with other U.S. industries, but with international industries.

The goal of any industry is to generate a profit. In today's marketplace this means creating the best product in the shortest time at a lesser price than the competition. A modern design process incorporating sophisticated analysis procedures and supported by high-speed computers with graphical displays increases the capability for developing the "best" product. The concept

Figure 1.38

Windmill farms are becoming an increasingly more important factor in the electrical infrastructure.

of integrating the design and manufacturing functions with CAD/CAM and CIM promises to shorten the design-to-market time for new products and for upgraded versions of existing products. The development of the automated factory is an exciting concept that is receiving a great deal of attention from manufacturing engineers today. Remaining competitive by producing at a lesser price requires a national effort involving labor, government, and distribution factors. In any case engineers are going to have a significant role in the future of our industrial sector.

1.11.4   Infrastructure

All societies depend on an infrastructure of transportation, waste disposal, and water distribution systems for the benefit of the population. In the United States much of the infrastructure is in a state of deterioration without sound plans for upgrading. For example:

1.   Commercial jet fleets include aircraft that are 30 to 35 years old. Major programs are now under way to extend safely the service life of these jets. In order to survive economically, airlines must balance new replacement jets with a program to keep older planes flying.

2.   One-half the sewage treatment plants cannot satisfactorily handle the demand.

3. The interstate highway system, nearing 45 years old in many areas, needs major repairs throughout. Local paved roads are deteriorating because of a lack of infrastructure funds.

4. Many bridges are potentially dangerous to the traffic loads on them.

5. Railroads continue to struggle with maintenance of the railbeds and the rolling stock in the face of stiff commercial competition from the air freight and truck transportation industries.

6. Municipal water systems require billions of dollars in repairs and upgrades to meet the demands of the public and the stiffening water quality requirements.

It is estimated that the total value of the public works facilities is over $2 trillion. To protect this investment, innovative thinking and creative funding must be fostered. Some of this is already occurring in road design and repair. For example, a method has been applied successfully that recycles asphalt pavement and actually produces a stronger product. Engineering research is producing extended-life pavement with new additives and structural designs. New, relatively inexpensive methods of strengthening old bridges have been used successfully.

1.11.5   Environment

Our insatiable demand for energy, water, and other national resources creates imbalances in nature which only time and serious conservation efforts can keep under control. The concern for environmental quality is focused on four areas: cleanup, compliance, conservation, and pollution prevention. Partnerships among industry, government, and consumers are working to establish guidelines and regulations in the gathering of raw materials, the manufacturing of consumer products, and the disposal of material at the end of its designed use.

The American Plastics Council publishes a document entitled *Designing for the Environment,* which describes environmental issues and initiatives affecting product design. All engineers need to be aware of these initiatives and how they apply in their particular industries:

- *Design for the environment (DFE).*   Incorporate environmental considerations into product designs to promote environmental stewardship.
- *Environmentally conscious manufacturing (ECM) or green manufacturing.*   Incorporate pollution prevention and toxics use reduction in the making of products to promote environmental stewardship.
- *Extended product or producer responsibility.*   Product manufacturers assume responsibility for taking back their products at the end of the product's life and disposing of them according to defined environmental criteria.
- *Life-cycle assessment (LCA).*   Quantified assessment of the environmental impacts associated with all phases of a product's life, often from the extraction of base minerals through the life of the product.
- *Pollution prevention.*   Reduce pollution sources in the design phase of products instead of address pollution after it is generated.
- *Toxic use reduction.*   Reduce the amount, toxicity, and number of toxic chemicals used in manufacturing.

As you can see from these initiatives, all engineers regardless of discipline must be environmentally conscious in their work. In the next few decades we will face tough decisions regarding our environment. Engineers will play a major role in making the correct decisions for our small, delicate world.

1.12 Conclusion

We have touched only briefly on the possibilities for exciting and rewarding work in all engineering areas. The first step is to obtain the knowledge during your college education that is necessary for your first technical position. After that you must continue your education, either formally by seeking an advanced degree or degrees, or informally through continuing education courses or appropriate reading to maintain pace with the technology, an absolute necessity for a professional. Many challenges await you. Prepare to meet them well.

Key Terms and Concepts

The following are some of the names, terms, and concepts you should recognize and understand.

Total quality	Recorder
Frederick Taylor	Team members
W. Edwards Deming	Team dynamics
Philip Crosby	Forming
Continuous incremental improvement	Storming
The customer	Norming
Quality control tools	Performing
Pareto chart	Team meetings
Cause-and-effect diagram	Consensus
The team	Personal style inventory
Culture change	Myers Briggs type indicator
Problem space	Introversion—extroversion
Solution space	Intuition—sensing
Team leader	Feeling—thinking
Facilitator	Perceiving—judging

Problems

1.1 Compare the definitions of an engineer and a scientist from at least three different sources.

1.2 Compare the definitions of an engineer and a technologist from at least two sources.

1.3 Find three textbooks that introduce the design process. Copy the steps in the process from each textbook. Note similarities and differences and write a paragraph describing your conclusions.

1.4 Find the name of a pioneer engineer in the field of your choice and write a brief paper on the accomplishments of this individual.

1.5 Select a specific branch of engineering and list at least 20 different industrial organizations that utilize engineers from this field.

1.6 Select a discipline of engineering, such as mechanical engineering, and an engineering function, such as design. Write a brief paper on some typical activities that are undertaken by the engineer performing the specified function. Sources of information can include books on engineering career opportunities, the World Wide Web (WWW), and practicing engineers in the particular discipline.

1.7 For a particular discipline of engineering, such as electrical engineering, find the program of study for the first 2 years and compare it with the program offered at your school approximately 20 years ago. Comment on the major differences.

1.8 Do Prob. 1.7 for the *last* 2 years of study in a particular branch of engineering.

1.9 List five of your own personal characteristics and compare that list with the list in Sec. 1.5.1.

1.10 Prepare a brief paper on the requirements for professional registration in your state. Include the type and content of the required examinations.

1.11 Prepare a 5-minute talk to present to your class describing one of the technical societies and how it can benefit you as a student.

1.12 Choose one of the following topics (or one suggested by your instructor) and write a paper that discusses technological changes that have occurred in this area in the past 15 years. Include commentary on the social and environmental impact of the changes and on new problems that may have arisen because of the changes.

(*a*) passenger automobiles
(*b*) electric power-generating plants
(*c*) computer graphics
(*d*) heart surgery
(*e*) heating systems (furnaces)
(*f*) microprocessors
(*g*) water treatment
(*h*) road paving (both concrete and asphalt)
(*i*) computer-controlled metal fabrication processes
(*j*) robotics
(*k*) air-conditioning

The following problems are intended to provide practice in the application of total quality principles.

1.13 *Class exercise.* The student counseling unit on your campus will provide a short overview of the Myers Briggs type indicator and allow students to complete the questionnaire.

1.14 *Class exercise.* The short version of the MBTI (Hogan and Champagne) will be distributed by your instructor (see ftn. 9). After completion and scoring, students will be placed in teams of three to five and asked to examine the description of all traits and to discuss the results of the inventory with others in the group. The team will prepare a brief report summarizing the collective team results and suggest any impact MBTI differences may have on this group working together over a period of time.

1.15 *Team projects.* You will be placed in a team of three to five students and either assigned or asked to select one of the customer improvement areas listed as follows. In general the total quality process consists of two parts: problem space and solution space. Your assignment will be to complete all or a portion of the problem space depending on the time available.

Possible Areas for Improvement

- Student parking on campus
- Transportation from residence to class locations
- Computer/printer access for students
- Used book market

- Team collaboration and study locations
- Living as a student on a tight budget
- Assigned by instructor
- Allocation of student activity fee resources
- Food service
- Area selected by student team.

Problem Space

Complete all the following steps or those assigned:

Step 1
- A. Identify customers.
- B. Develop interview questions.
- C. Interview customers.

Step 2
- D. Sort customer input.
  Affinity diagram
- E. Quantify customer concerns.
  Pareto chart
- F. Issue abatement.

Step 3
- G. Diagram the process.
  Flowchart

Step 4
- H. Diagram causes and effects.
  Fishbone diagram

Step 5
- I. Collect and analyze data on causes.

Complete a team report according to guidelines provided by your instructor.

CHAPTER 2

Engineering Design—A Process

2.1   Introduction

Engineering design is a systematic process by which solutions to the needs of humankind are obtained. Design is the essence of engineering. The design process is applied to problems (needs) of varying complexity. For example, mechanical engineers will apply the design process to develop an effective, efficient lawn mulching machine; electrical engineers will apply the process to design lightweight, compact wireless communication devices; and materials engineers will apply the process to design high-temperature materials which protect astronauts during re-entry in the earth's atmosphere.

The vast majority of complex problems in today's high-technology society do not depend for solutions on a single engineering discipline; rather, they depend on teams of engineers, scientists, environmentalists, economists, sociologists, legal personnel, and others. Solutions are not only dependent on the appropriate applications of technology but also on public sentiment as executed through government regulations and political influence. As engineers we are empowered with the technical expertise to develop new and improved products and systems; however, at the same time we must be increasingly aware of the impact of our actions on society and the environment in general and work conscientiously toward the best solution in view of all relevant factors.

A formal definition of *engineering design* is found in the curriculum guidelines of the Accreditation Board for Engineering and Technology (ABET). ABET accredits curricula in engineering schools and derives its membership from the various engineering professional societies. Each accredited curriculum has a well-defined design component which falls within the ABET guidelines. The ABET statement on design reads as follows:

> Engineering design is the process of devising a system, component, or process to meet desired needs. It is a decision-making process (often iterative), in which the basic sciences, mathematics, and engineering sciences are applied to convert resources optimally to meet a stated objective. Among the fundamental elements of the design process are the establishment of objectives and criteria, synthesis, analysis, construction, testing, and evaluation. The engineering design component of a curriculum must include most of the following features: development of student creativity, use of open-ended problems, development and use of modern design theory and methodology, formulation of design problem statements and specifications, consideration of alternative solutions, feasibility considerations, production processes, concurrent engineering

design, and detailed system descriptions. Further, it is essential to include a variety of realistic constraints such as economic factors, safety, reliability, aesthetics, ethics, and social impact.

Figure 2.1

1. Identification of a need
2. Problem definition
3. Search
4. Constraints
5. Criteria
6. Alternative solutions
7. Analysis
8. Decision
9. Specification
10. Communication

A 10-step design process.

In order to gain the fundamental knowledge and experience needed to understand the design process, you must partake in meaningful design activities as a student. To assist you in your first activity, we will take you through a design project appropriate for a team of beginning engineering students. The project is guided by the 10-step design process listed in Fig. 2.1. As you study the design process and the description of how a student team might accomplish each step, pay particular attention to the structure of the process rather than to the development of the particular solution arrived at by the team. By doing this you will understand how engineers approach a need, develop alternative solutions, select the best solution, and communicate the results.

Throughout this chapter you will see examples of the utilization of computers in engineering design (see Fig. 2.2). The computer is a major tool for the engineer in the acquisition and analysis of data, and the definition of potential solutions. Throughout history engineers have used the best computational devices available at the time to obtain solutions to problems. The computer performs numerical computations at the rate of billions per second. It also provides an insight to problems and solutions through its capability to simulate actual phenomena. This capability provides the engineer with a tremendous advantage in developing new and improved products in a much shorter time frame than ever before. The World Wide Web (WWW) provides almost instant access to information previously obtained through time-consuming library research, searching company documents, consulting with experts in the field, and so on.

2.1.1   The Design Process

A simple definition of *design* is "a structured problem-solving activity." A *process,* on the other hand, is a phenomenon identified through step-by-step changes that lead toward a required result. Both these definitions suggest the idea of an orderly, systematic approach to a desired end. Figure 2.3 shows the design process approach as continuous and cyclic in nature. The design process, however, is not linear. That is, one does not necessarily achieve the

Figure 2.2

A design team member generating a logic circuit for a computer chip.

Figure 2.3

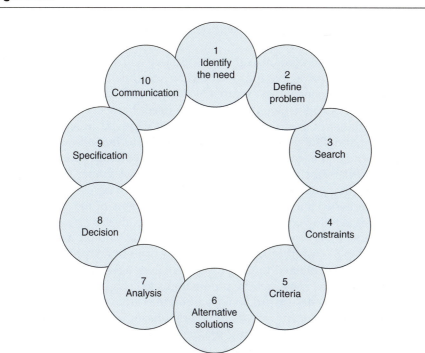

The design process is iterative in nature.

Figure 2.4

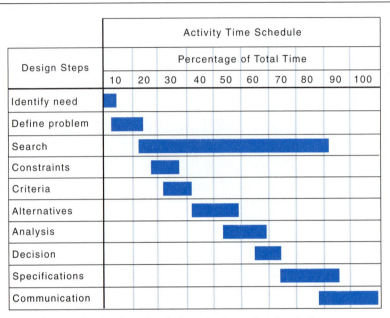

	Activity Time Schedule									
Design Steps	Percentage of Total Time									
	10	20	30	40	50	60	70	80	90	100
Identify need	▮									
Define problem	▮▮									
Search		▮▮▮▮▮▮▮								
Constraints			▮							
Criteria			▮							
Alternatives					▮					
Analysis						▮				
Decision							▮			
Specifications								▮▮		
Communication									▮▮	

A time schedule must be developed early in order to control the design process.

best solution by simply proceeding from one step in the process to the next. New discoveries, additional data, and previous experience with similar problems generally will result in several iterations through some or all the steps of the process. The chapter example will clearly illustrate the iterative nature of design.

It is important to recognize that any project will have time constraints. Normally before a project is approved a time schedule and a budget will be approved by management. An example of a time schedule is shown in Fig. 2.4.

The process begins when a need is recognized: In essence, a human need has been identified. Often the designers are not involved in step 1; but they usually assist in defining the problem (step 2) in terms that allow it to be scrutinized. Information is gathered in step 3, and then boundary conditions (constraints) are established (step 4). The criteria against which the alternative solutions are compared are chosen in step 5. At step 6 several possible solutions are entertained and the creative, innovative talents of the designer come into play. This is followed by detailed analysis of the alternatives (step 7), after which a decision is reached regarding which one should be completely developed (step 8). Specifications of the chosen concept are prepared (step 9), and its merits are explained to the proper people or agencies (step 10) so that implementation (e.g., construction, production, etc.) can be accomplished.

The solution to a design problem, as accomplished by iteration through the 10 previously defined steps, occurs in several phases in most industrial situations. See Fig. 2.5. After the identification and definition of a problem (design steps 1 and 2), the conceptual design phase begins. At this time information is gathered (design step 3), constraints are established (design step 4), and a

Figure 2.5

83

Introduction

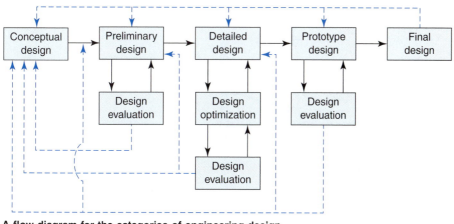

A flow diagram for the categories of engineering design.

multitude of possible solutions are generated (design step 6). There are no decisions made on the merit of any of the possible solutions. The preliminary design phase occurs next and includes the development of the criteria (design step 5) and the analysis of potential solutions (design step 6 results are modified here before analysis takes place). As you will see in the chapter example, the analysis of potential solutions (design step 7) provides the basis for determining the merit of each solution with respect to the solution criteria. The conclusion of the preliminary design phase is the selection of the best solution (design step 8).

The detailed design phase includes the selection of stock parts, the design of all other components, and the optimization of the solution (design step 9). Optimization includes consideration of cost, materials, performance, manufacturability, and feasibility. It is important to note that the criteria (design step 5) are used to select the best solution from the list of possible solutions. *Optimization* is the process of determining the best solution based on an evaluation of a "short" list of final solution candidates. As an example, consider the selection of a transportation system for moving goods from the East Coast of the United States to the West Coast. Possible solutions include trucks, trains, ships, and airplanes. Assume that, based on the criteria in design step 5, airplanes become the selected mover. What airplane or airplanes are designed or selected for the task? It is likely that the *optimum* airplane would be the one that transports the goods for the least cost per pound.

The next phase (Fig. 2.5) is the prototype design. The development of a new product usually involves a prototype and extensive testing before mass production begins. The information flow lines on Fig. 2.5 (solid lines) and the feedback lines (dashed) again illustrate the iterative nature of the design process. Without this continuous evaluation during the process, it is not possible to obtain an optimum solution. The last phase is the final design, which is communicated (design step 10) to manufacturing for production.

Figure 2.6

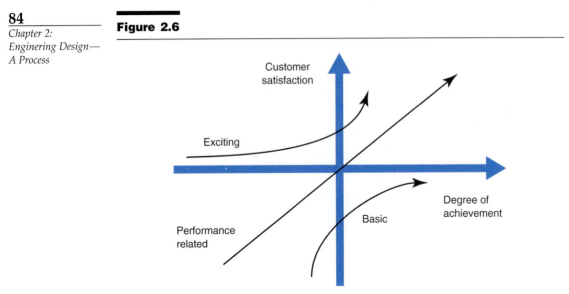

Factors in generating customer satisfaction.

2.1.2   Design and the Customer

The result of the execution of the design process is a new product, process, or system. The process started with the identification of a need which likely was suggested or requested by a customer. Thus the final design must satisfy customer requirements or the design may be rejected. Often customer requirements are not well defined or are vague. The design team must determine, in consultation with the customer, the expectations of the solution. The customer therefore must be kept informed of the design status at all times during the process. It is likely that compromises will have to be made. Both the design team and customer may have to modify their requirements in order to meet time deadlines, cost limits, manufacturing constraints, and performance requirements. Figure 2.6 is a simple illustration of the Kano model showing the relationship between degree of achievement (horizontal axis) and customer satisfaction (vertical axis). Customer requirements are categorized in three areas: basic, performance related, and exciting.

Basic customer requirements are simply expected by the customer and assumed to be available. For example, if the customer desires a new electric-powered barbeque grill, the assumption is made that the design team and the company have proven their ability, with existing successful products, to design and manufacture electric-powered barbeque grills.

Performance-related customer requirements are the basis for requesting the new product. In the example of the barbecue grill, cooking time, cooking effectiveness, ease of setting the controls, ease of cleaning are among the many possible performance-related items that a customer may specify. As time goes by and more electric-powered grills reach the marketplace, these requirements may become basic.

Exciting customer requirements are generally suggested by the design team. The customer is unlikely to request these features because they are often outside the range of customer knowledge or vision. The exciting requirements are often a strong selling point in the design because they give the customer an unexpected bonus in the solution. Perhaps the capability of programming a cooking cycle to vary the temperature during the cooking process would be a unique (but perhaps costly) addition to the solution.

Figure 2.6 indicates that the basic requirements are a must for customer satisfaction. The customer will be satisfied once a significant level of performance-related requirements are met. The exciting requirements always add to customer satisfaction, so the more of these features that can be added, the better is the satisfaction.

2.1.3 The Nature of Engineering Design

Until recent years engineering design has been considered by many to be a creative, ad hoc process which did not have a scientific basis. Design was considered an art, with successful designs emanating from a few talented individuals in the same manner as great artwork is produced by talented artists. However, there are now convincing arguments that engineering design is a cognitive process that requires a broad knowledge base, intelligent use of information, and logical thinking. Today successful designs are generated by design teams, comprised of engineers, marketing personnel, economists, management, customers, and so on, working in a structured environment and following a systematic strategy. Utilizing tools such as the WWW, company design documentation, brainstorming, total quality management (TQM), and the synergy of the design team, information is gathered, analyzed, and synthesized with the design process yielding a final solution that meets the design criteria.

What do we mean by a cognitive process? In the 1950s Benjamin Bloom developed a classification scheme for cognitive ability that is called Bloom's taxonomy. Figure 2.7 shows the six levels of complexity of cognitive thinking and provides an insight into how the design process is an effective method of producing successful products, processes, and systems. The least complex level,

Figure 2.7

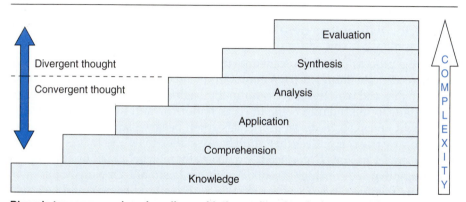

Bloom's taxonomy on learning aligns with the engineering design process.

knowledge, is simply the ability to recall information, facts, or theories. [What was the date of the *Challenger* space shuttle accident?]

The next level is comprehension, which describes the ability to make sense (understand) the material. [Explain the cause of the *Challenger* accident.] The third level is application, which is the ability to use knowledge and comprehension in a new situation and to generalize the knowledge. [What would you have done to prevent the *Challenger* accident?]

The fourth level is analysis, which is the ability to break learned material into its component parts so that the overall structure may be understood. It includes part identification, relationships of the parts to each other and to the whole, and recognition of the organizational principles involved. The individual must understand both the content and structure of the material. Figure 2.7 shows that analysis is the highest level of convergent thinking, whereby the individual recalls and focuses on what is known and comprehended to solve a problem through application and analysis. [What lessons did we learn about the space program from the *Challenger* accident?]

Levels 5 and 6 on Bloom's taxonomy represent divergent thinking, in which the individual processes information and produces new insights and discoveries that were not part of the original information. Synthesis refers to the ability to put parts together to form a new plan or idea. Everyone synthesizes in a different manner. Some accomplish synthesis by quiet mental musing; others must use pencil and paper to doodle, sketch, outline ideas, and so on. [Propose an alternative to the *Challenger* O-Ring design that would perform the required functions.]

Evaluation is the highest level of thinking. It is the ability to judge the value of material based on specific criteria. Usually the individual is responsible for formulating the criteria to be used in the evaluation. [Assess the impact of the *Challenger* accident on the U.S. space program.]

To help your understanding of the levels of cognitive thinking, review several exams you have taken in college in mathematics, chemistry, physics, and sociohumanistic courses (e.g., economics, sociology, history, etc). Take each question and decide which level of thinking was required to obtain a successful result. You will find while moving along in your engineering curriculum that exam questions, homework problems, and projects will reflect higher and higher levels of thinking.

2.2 Identification of a Need—Step 1

Before the process can begin someone has to recognize that some constructive action needs to be initiated. This may sound vague, but understandably so because such is the way the process normally begins. Engineers do not have supervisors who tell them to come to work tomorrow and to "identify a need." You might be asked to do so in the classroom because some professors may have you work on a project that you choose rather than one that is assigned. When most of us speak of a need, we generally refer to a lack or shortage of something that we consider essential or highly desirable. Obviously this is an extremely relative thing, for what may be a necessity to some could be a luxury to others.

More often than not, then, someone other than the engineer decides that a need exists. In industry, or in the private sector, it is essential that products sell for the company to remain in business. Most products have a life cycle that goes from the development stage, when the expenditures by the company are high and sales are low, to the peak demand period, when profits are high, and eventually to the point where the product becomes obsolete. Even though a human need may still exist, the economic demand does not because a more attractive alternative has become available. With obsolescence of a product, the company perceives a need to phase out the existing product and to develop a new one that is profitable. Inasmuch as most companies exist to make a profit, profit can be considered to be the basic need.

A bias toward profit and economic advantage should not be viewed as a selfish position because products are purchased by people who feel that what they are buying will satisfy a need that they perceive as real. Society appreciates anyone who provides essential and desirable services, as well as goods that we use and enjoy. The consumers are ultimately the judges of whether there is truly a need. In like manner, the citizens of a community decide whether or not to have paved streets, parks, libraries, adequate police and fire protection, and scores of other things. City councils vote on the details of the programs. However, during the period when citizens and decision makers are formulating their plans, engineers are involved in supplying factual information to assist them. After the policy decisions have been made, engineers conduct studies, surveys, tests, and computations that allow them to prepare the detailed design plans, drawings, and so on that shape the final project.

2.2.1 The Chapter Example—Step 1 of the Design Process

The best way to develop a capability to perform engineering design is to go through the design process and to arrive at a solution to a real problem. The remainder of the chapter will illustrate how the design process might be applied by a small team (three to five members) of beginning engineering students to a problem which has a direct connection to the quality of life on a campus. The material presented here is based on experiences and conditions at Iowa State University and may or may not reflect conditions at other colleges and universities. As you go through the example, focus on the steps of the design process that lead to the final solution, but at the same time reflect on the situation at your campus and note how you would have handled this problem in your environment. Keep in mind that the solution to a design problem is a result of the information available, the makeup of the design team, and the decisions that are made at each step of the process. There is not one right answer, but several possibilities that satisfy the problem definition and criteria. It is your task, as a design team, to propose the best solution in the time you are allowed to work on the problem.

Dormitories in general are designed to accommodate two students and normally are equipped with two desks, two closets, two beds, two shelving units, and one dresser. Needless to say, the room is very crowded and if a refrigerator, microwave oven, lounge chair, and television are permitted, open space is very limited. If the room occupants also have computers, storage for books, papers, and other school materials is quite limited. One modification that frees up space is to loft or elevate the beds. This gives access to the area under the bed but creates a new problem of how to arrange the other furniture to take advantage of the freed space. In addition, lofting the bed possibly creates other problems such as lack of stability, high cost, and difficulty in assembling and disassembling. A combined system that includes a lofted bed, closet, desk, dresser, and shelves would seem to be an ideal solution to the space problem. A common design specified by the college or university for all its dormitories would be beneficial.

There is a need for a system that will utilize better dormitory living space for college and university students.

2.3 Problem Definition—Step 2

There is often a temptation to construct quickly a mental picture of a gadget that if properly designed and manufactured will satisfy a need. In the general case of a need for a loft system it is not a difficult task to come up with a system that "works." A few boards, some nails, and a hammer might produce a satisfactory solution for some students. How about stability? Efficiency of assembly? Efficient use of space? Existing university regulations on what may be modified in a dormitory room? These and many other questions indicate that more thought and care needs to go into the loft design. Thus before we think solutions let us be sure that the problem we want to solve is well defined.

2.3.1 Broad Definition First

The need as previously stated does not point to any particular solution and thereby leaves us with the opportunity to consider a wide range of alternatives before we agree on a specific problem statement. Consider for a moment a partial array of possibilities that will satisfy the original statement that there is a need for a system that will better utilize dormitory living space.

1. Purchase a prefabricated loft system.
2. Sketch a possible solution on a piece of paper, obtain approval, purchase parts, and assemble.
3. Rent two rooms and cut a connecting door.
4. Purchase an existing loft system from a graduating senior.
5. Use the engineering design process to find the best system according to specified criteria.

Another possibility is to use the room as furnished and to locate the furniture to provide the maximum free space. This is the status quo and often is the solution chosen, particularly on a temporary basis. The solution listed third, while satisfying the need, is not practical. Thus we see that a careful problem definition is necessary before proceeding to alternative solutions.

2.3.2 Symptom Versus Cause

If you cough and only suck on cough drops, you may be treating the symptom (the tickle in your throat) but doing little to alleviate the cause of the tickle.

This approach may be expedient; however, it can result many times in a repetition of the problem if the tickle is caused by a virus or a foreign object of some sort. Engineers seldom tell a client to take two aspirins and call back tomorrow, but they can sometimes be guilty of failing to see the real or root problem.

For many years residential subdivisions were designed so that the rainfall would drain away quickly, and expensive storm sewer systems were constructed to accomplish this task. Not only were the sewers expensive, but they also resulted in transporting the water problem downstream for someone else to handle. In recent years perceptive engineers have designed land developments so that the rainfall is temporarily collected in "holding pools" and released gradually over a longer period of time. This approach employs smaller, less expensive sewers and reduces the likelihood of flooding downstream. The real problem was not how to get rid of the rainfall as rapidly as possible, but how to control the water.

2.3.3 Solving the Wrong Problem

In the 1970s the problem of increasing fatalities as a result of auto accidents was clearly recognized. It was shown that the fatality rate could be reduced significantly if the driver and front seat passenger used lap and shoulder belts. The solution technique that was implemented was to build in an interlock system that required the belts to be latched before the auto could be started. That solution certainly should have solved the problem but it did not. It attacked the problem of requiring that the belts be physically used; however, it did nothing to solve the real problem—the driver's attitude. The driver and passenger still did not wish to use belts and did everything possible to avoid it, even by having the interlock system removed.

2.3.4 The Chapter Example—Step 2

To obtain the most concise and complete problem definition based on the perceived need, we will use the state A → state B designation, where state A represents the undesirable situation and state B represents the desirable situation. Using this approach, we can be assured that we are solving the correct problem.

For example, consider the following:

Crowded living conditions → uncrowded living conditions

This is a very broad definition of the problem and permits solutions such as renting or buying a house, renting a large apartment, or paying double rent so you can have a dorm room to yourself (if this were permitted). The possible solutions for this problem are numerous and would not achieve the result we perceive from the need statement.

The next step is to redefine the problem statement by narrowing the solution possibilities to dormitories. A possible problem statement is

Existing dorm furnishings → existing furnishings with lofted
beds

This statement implies that the bed loft would be designed to accommodate existing desks and shelves. If these furnishings are not standard in size, then several loft design modifications may be needed. In view of the time allotted for the design project, we can reasonably reduce the scope of the problem such that only the beds are designed. It certainly seems appropriate to try and design a complete dormitory setup for good utilization of space. However, considering the time constraint of the academic environment in which the student design teams must work, we can reasonably reduce the scope so that a final design is achievable within the objectives of the learning experience.

The final problem statement becomes

Existing dorm beds → lofted beds

This problem definition restricts the solution to one involving lofted beds, which fits the stated need more closely. It permits a wide range of possible solutions, namely designs that involve different arrangements of the components. There is great latitude available for student preferences and needs within university constraints. At this point a good engineering design team will reflect on the customer's needs to make sure that the problem definition meets the initial customer's requirements. This is a good time to meet again with the customers (students) to make sure everyone is "on the same page."

2.4 Search—Step 3

Most of your productive professional time will be spent locating, applying, and transferring information—all sorts of information. This is not the popular opinion of what engineers do, but this is how it will be for you. Engineers are problem solvers, skilled in applied mathematics and science, but they seldom, if ever, have enough information about a problem to begin solving it without first gathering more data. This search for information may reveal facts about the situation that result in redefinition of the problem. Remember, at this point no formal list of possible solutions has been developed. We must continue to think in terms of the problem at hand and how we can increase our knowledge, which will in turn lead us to a better solution.

2.4.1 Types of Information

The problem usually dictates what types of data are going to be required. The one who recognizes that something was needed (design step 1) probably listed some things that are known and those things that need to be known. The one or ones who defined the problem had to have knowledge of the topic or they could not have done their part (design step 2). Generally we look for several things when beginning to solve most problems. For example;

1. What has been written about it?
2. Is something already on the market that may solve the problem?
3. What is wrong with the way it is being done?
4. What is right with the way it is being done?
5. Who manufactures the current "solution"?

6. How much does it cost?
7. Will people pay for a better one if it costs more?
8. How much will they pay (or how bad is the problem)?

2.4.2 Sources of Information

If anything can be said about the last half of the twentieth century, it is that we have had an explosion of information. The amount of data that can be uncovered on most subjects is overwhelming. People in the upper levels of most organizations have assistants who condense most of the things that they must read, hear, or watch. When you begin a search for information, be prepared to scan many of your sources and to document their location so that you can find them easily if the data subsequently appear to be important.

Some sources that are available include the following, although not all of these will be appropriate sources for our particular problem.

1. *Existing solutions.*   Much can be learned from the current status of solutions to a specific need if actual products can be located, studied, and, in some cases, purchased for detailed analysis. If a product can be acquired in the marketplace, a process called *reverse engineering* can be performed to determine answers to some questions posed in the previous section. An improved solution or an innovative new solution cannot be found unless the existing solutions are thoroughly understood. Reverse engineering is an excellent learning technique for students and engineers in industry who are beginning to apply the design process. You should look to local industries and retail outlets for existing solutions that currently satisfy the need you have identified. In some instances you may be able to purchase the product or at least to observe a demonstration of its capability.

2. *Internet.*   The electronic Internet that today connects millions of computer users in nearly 100 nations to a global "information superhighway" provides rapid access to a wealth of knowledge. Industries, businesses, government organizations, and educational institutions are now connected to the Internet enabling nearly instantaneous access to research data, product information, special interest groups, and experts throughout the world. Although you may not yet have access to the Internet from your personal computer, your university has facilities from which you can gain access. Accessing the Internet for information should be a major component in any research effort. Divide the WWW search (key word categories, company homepages, etc.) among the team members for an effective and efficient search. Then hold a team meeting after initial efforts, compare findings, and decide how to continue the search (if necessary).

3. *Your library.*   Many universities have courses that teach you how to use your library. Such courses are easy when you compare them with those in chemistry and calculus, but their importance should not be underestimated. There are many sources in the library that can lead to the information that you are seeking. You may find what you need in an index, such as the *Engineering Index,* but do not overlook the possibility that a general index, such as *The Reader's Guide* or *Business Periodicals Index,* also may be useful. The *Thomas Register of American Manufacturers* may direct you to a company that makes a product about which you need to know more. *Sweets Catalog* is a compilation

of manufacturers' information sheets and advertising material about a wide range of products. There are many other indexes that provide specialized information. The nature of your problem will direct which ones may be helpful to you. Do not hesitate to ask for assistance from the librarian. It is to your advantage to use the computer databases that are found in libraries and often available through CD-ROM technology.

4. *Government documents.* Many of these are housed in special sections of your library, but others are kept in centers of government—city halls, county court houses, state capitols, and Washington, DC. The regulatory agencies of government, such as the Interstate Commerce Commission, the Environmental Protection Agency, and regional planning agencies, make rules and police them. The nature of the problem will dictate which of the myriad of agencies can fill your needs.

5. *Professional organizations.* The American Society of Civil Engineers (ASCE) is a technical society that will be of interest to students majoring in civil engineering. Each major in your college is associated with not one but often several such societies. The National Society of Professional Engineers (NSPE) is an organization that most engineering students will eventually join, as well as at least one technical society such as the American Society of Mechanical Engineers (ASME), the Institute of Electrical and Electronics Engineers (IEEE), or any one of dozens that serve the technical interests of the host of specialties with which professional practices seem most closely associated. Many engineers are members of several associations and societies. Other organizations, such as the American Medical Association (AMA) and the American Bar Association (ABA), serve various professions, and all have publications and referral services.

6. *Trade journals.* These are published by the hundreds, usually specializing in certain classes of products and services.

7. *Vendor catalogs.* Perhaps your college subscribes to one of the several information services that gather and index journals and catalogs. These data banks may have tens of thousands of such items available to you on microfilm. You only need to learn how to use them.

8. *Individuals whom you believe are experts in the field.* Your college faculty has at least several such individuals, maybe many. There are, no doubt, some practicing engineers in your city.

2.4.3 Recording Your Findings

The purpose of a bibliography is to direct you to more information than that included in the article you are reading. The form of the bibliography makes it easy to find the reference, thus enabling you to record your information sources in proper form so that if you cite the reference in your report, you are ready to do it properly. By so doing, you are ensuring that it can be found again quickly and easily. Few things are more discouraging than being unable to locate an article that you found once and know it will be helpful if you could locate it again.

Usually a good procedure is to record each reference on a card or sheet of paper or to maintain a computer database. English teachers usually recommend the use of file cards, but engineers seem to prefer information put in a bound

notebook or in the computer. You should check for the proper format for reporting book sources, technical papers, documentation of personal interviews,
and WWW sources.

2.4.4   The Chapter Example—Step 3

Based on the efforts of several beginning engineering student design teams at
Iowa State university, the search focused on four areas: university restrictions
and specifications, existing solutions, student preferences, and construction
materials.

The university restrictions and specifications were obtained through personal interviews with residence hall officials and are summarized here:

- Loft system must meet current Iowa State, City of Ames, and State of
  Iowa safety codes.
- Loft system must be able to be assembled with simple tools.
- Loft system should have as many standard parts as possible.
- Loft system does not need a guard rail or ladder.
- Loft system does not need to be handicap accessible.
- Students have indicated a preference for wood construction.

Several websites were investigated to determine designs, cost, structure and
materials. A few URLs of these websites are given here. The reader is encour-
aged to access these sites and to note the information that is available.

- http://detnews.com/menu/storie/13826.htm
- http://www.durabull.com/wizzm.html
- http://www.ecoloft.com/benefits.html
- http://www.loftbeds.com

In addition to the websites, at least one lumber yard in Ames builds a standard
loft for dormitory students. Costs were found to vary considerably. By purchasing the parts and doing all the assembly, students can build a loft for less
than $100. Some commercially produced lofts, which include desks and shelves,
cost as much as $1,500.

Student preferences were obtained for the most part through surveys and
interviews. The design teams made sure a reasonable cross-section of students
were surveyed to ensure good results. Upper-division students who had used
a loft system and beginning students who had not yet used a loft system were
surveyed. Figure 2.8 shows a typical survey and the key results obtained.

Possible materials for the loft system were looked into by checking lumber yards and welding shops. Although steel would provide a strong, stable
loft system, working with metal (i.e., for drilling, cutting, and filing) was determined to be much more difficult than working with wood. Thus most design teams selected wood as the construction material. Costs for basic materials such as oak hardwood, oak veneer plywood, wood stain and finishing
products, and metal fasteners were compiled. The material cost for a specific
loft design can be quickly computed from the known component costs.

For the dormitories at Iowa State, cordless screwdrivers, hammers,
wrenches, power saws, and hand saws are available for checkout. There would
be no added cost for tools for the loft construction.

Figure 2.8

Dormitory Loft Survey

1. Do you prefer a loft system to the existing dorm room furnishings? Yes 86% No 14%

2. Would you prefer to have a loft already installed in your room? Yes 18% No 82%

3. Give reasons for your answer to question 2. —Want to customize room (no) —Don't want to be forced to use a particular loft design (no)—I don't have time to design and install a loft (yes)—

4. What would you be willing to pay for a loft system?

70%	< $100
15%	$100 – $250
10%	$250 – $500
5%	> $500

5. Rate on a scale of 1 to 10, with 10 the most important, each of the following characteristics of a loft system:

8	Durability
7	Accessibility
6	Stability
9	Cost
4	Appearance
10	Ease of assembly
8	Safety
5	Maintenance

A summary of results of the student survey on dormitory lofts. Documentation of research findings is essential if the findings are to be useful later.

2.5 Constraints—Step 4

Up to this point we have kept the problem definition as broad as possible, within the scope of available time and the level of expertise of the design team, but still allowing for a large number of potential solutions. However, there are physical and practical limitations (called *constraints*) that will reduce the number of solutions for any problem. For example, in order to remain competitive in the marketplace, a retail cost of $50 for a particular solution cannot be exceeded. Another example is that any conceived solution must be able to operate using a standard household 120 V (volt) outlet. In many cases the physical size of the solution is limited by market competition or legal restrictions. The laptop computer is a good example of placing size (and weight) constraints on a solution in order to meet the competition in the marketplace. Whenever a *constraint* is applied, the solution possibilities are reduced; therefore you should take care that the constraint is not "artificial," that is, not overly restrictive to the process of finding an innovative solution.

We face a similar situation in almost every decision we make, even though sometimes the decisions are not really important. When arising this morning, you had to choose some clothes to wear. You probably limited the choice to those hanging in your closet (or maybe in your roommate's closet). This was a constraint *you* placed on the situation, not one that really existed until you made it so. In most fields of engineering formulas have been developed and are used in designs of various kinds. Many, and probably most, of them are valid in a certain range of physical conditions. For instance, the hydraulic conditions of the flow of water are not valid below 0°C or above 100°C and are restricted to normal pressure ranges. We normally refer to these constraints or limits as *boundary conditions*, and they occur in many different ways.

2.5.1 The Chapter Example—Step 4

The results of the search step revealed several things that limit or eliminate potential solutions. Before generating alternative solutions, we list the following constraints:

- Cost must not exceed $1500.
- Loft system must meet safety and fire codes for Iowa State, City of Ames, and the State of Iowa.
- Loft system must accommodate a unit bed size of 78 × 36 inches.
- Loft system must be freestanding and cannot affect the existing structure of the room.

2.6 Criteria—Step 5

Criteria are desirable characteristics of the solution which are established from experience, research, market studies, and customer preferences. In most instances the criteria are used to judge alternative solutions on a qualitative basis. However, if a performance function can be defined mathematically, then an optimum solution can be obtained numerically. The mathematical method of optimization is beyond the scope of this introductory text. Instead, we concentrate on the selection and weighting of a set of criteria that will produce the best solution for the stated problem and solution constraints.

2.6.1 Design Criteria

Whereas each project or problem has a personality all its own, there are certain characteristics that occur in one form or another in a great many projects. We should ask ourselves, "What characteristics are most desirable and which are not applicable?" A list of typical design criteria follows:

1. Cost—almost always a heavily weighted factor
2. Reliability
3. Weight (either light or heavy)
4. Ease of operation and maintenance
5. Appearance
6. Compatibility

Figure 2.9

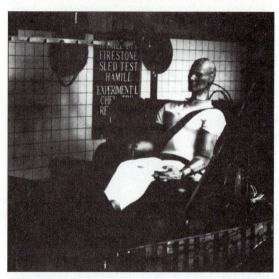

The simulation of accidents on a computer system assists engineers in the establishment of design criteria for safer vehicles. (*Courtesy of the ISU Visualization Laboratory.*)

7. Safety features
8. Noise level
9. Effectiveness
10. Durability
11. Feasibility
12. Acceptance

There will be other criteria, and perhaps some of those given are of little or no importance in some projects, so a design team in industry or in the classroom must decide which *criteria* are important to the design effort (see Fig. 2.9). Since value judgments have to be made later, it probably makes little sense to include those criteria that will be given relatively low weights. There are often mild disagreements at this point, not about which criteria are valid but, rather, about how much weight should be assigned to each. It is frequently better if the team members make their assignments of weight independently and then compile all the results. This tends to dampen the effect of the more persuasive members at the same time that it forces all team members to contribute consciously. Usually there are not many instances where one of the members strongly disagrees with the mean value of the weight assigned to each criterion. Some negotiation may be required, but it is seldom a difficult situation to resolve.

Perhaps one way to resolve any differences of opinion on the selection and weight of criteria is carefully to define each criterion. This will generally clear up differences because two or more criteria may be combined into one. For example, reliability and durability could be combined into dependability.

Figure 2.10

97
Criteria—Step 5

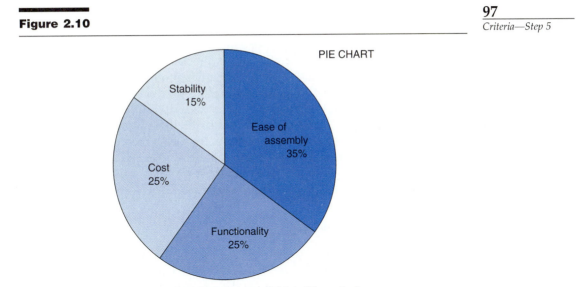

A visual chart helps to understand the relative weights of the criteria.

2.6.2 The Chapter Example—Step 5

With the results of the search phase as a guide, the design team must now arrive at a list of criteria from which a final solution will be determined by evaluating the alternatives against the criteria. For purposes of illustration the following criteria and weights are presented (see Fig. 2.10). These vary considerably among design teams depending on the interpretation of search results and the opinions of the design team members.

1. *Ease of assembly (35%).* This criterion is self-explanatory and has a high weight based on the results of the student survey.
2. *Functionality (25%).* This includes accessibility to the loft, the effective use of space, durability, and maintenance.
3. *Cost (25%).* This includes the cost of materials and tools.
4. *Stability (15%).* This includes consideration of structural integrity, sense of safeness, and capability of being freestanding.

The astute reader has noticed that cost can be both a constraint and a criterion. This is typical in designs that are competing with similar designs in the marketplace. As a constraint, cost is given as an upper limit. For example, a 35 mm (millimeter) camera with a zoom lens may have a design constraint that the cost will not exceed $300. However, when considering alternative solutions or different brands, you may select one for $259 over one priced at $279, provided the performance characteristics are similar. Both cameras meet the design constraint but the $20 difference would give the lower-priced camera the advantage in the marketplace.

The process of selecting the best solution from the alternative list will show that having a short list of criteria is superior to a list containing many criteria with low weights. In this case four criteria were presented; normally four to six is a reasonable number of criteria.

2.7 Alternative Solutions—Step 6

Suppose that you are chief engineer for a manufacturing company and are faced with appointing someone to the position of director of product testing. This is an important position because all the company's products are given rigorous testing under this person's direction before the products are approved. You must compare all the candidates with the job description (criteria) to see who would do the best job. This seems to be a ridiculously simple procedure, doesn't it? Well, we think so too, but many times such a process is not followed and poor appointments are made. In the same way that the list of candidates for the position has to be made, thus we must produce a list of possible answers to our problem before selecting the best one.

2.7.1 The Nature of Invention

The word "invention" strikes fear into the minds of many people. They say, "Me, an inventor?" The answer is "Why not?" One reason that we do not fashion ourselves as inventors is that some of our earliest teaching directed us to be like other children. Since much of our learning was by watching others, we learned how to conform. We also learned that if we were like the other kids, no one laughed at us. We can recall in our early days, even preschool, that the worst thing that could happen would be if people laughed at us. We bet that most of you have a similar feeling when saying something that is not too astute, and it is followed by smiles and polite laughs. Moreover, we do not like to experiment because many experiments fail. Only a very secure person never has to try something that he or she has not already done well. Think about it: When in the first few grades at school, didn't you feel great when you were called on by the teacher and you knew the answer? Don't you, even today, try to avoid asking your professors a question because you do not want the professor or your classmates to know that you do not know the answer? Most of us like to be in the majority. Please do not assume that we are saying that the majority of the people are wrong or that it is bad to be like other people. However, if we dwell on such behavior, then we will never do anything new. A degree of inventiveness or creativity is essential if we are to arrive at solutions to problems that are better than the way things are being done now. If we can remove the blocks to creativity, then we have a good chance of being inventive.

2.7.2 Building the List

There are numerous techniques that can be used to assist us in developing a list of possible solutions. Two of the more effective methods will be briefly discussed.

Checkoff lists, designed to direct your thinking, have been developed by a number of people. Generally the lists suggest possible ways that an existing solution to your problem might be changed and used. Can it be made a different color, a different shape, stronger or weaker, larger or smaller, longer or shorter, of a different material, reversed or combined with something else, and so forth? It is suggested that you write your list down on paper and try to conceive of how the current solution to the problem might be if you changed it according to each of the words on your list. Ask yourself: Why is the solution like it is? Will change make it better or worse? Did the original designers have good reason for doing what they did or did they simply follow the lead of their predecessors?

Use the checklist words "modify" and "rearrange" to guide or focus on the efforts to obtain new solutions for the loft design:

Modify?   Use laminates instead of solid wood.
           Use glue instead of bolts or screws.
Rearrange? Bed on the floor, desk, and shelves above.
           Ladder on side of bed instead of end.

Brainstorming is a technique that has received wide discussion and support. The mechanics of a brainstorming session are rather simple. The leader states the problem clearly and ideas about its solution are invited. The length of productive sessions varies, but it is usually in the one-half hour range. Often a group takes a few minutes to rid itself of its natural reserved attitude. But brainstorming can be fun, so choose a problem area and try it with some friends. Be prepared for a surprise at the number of ideas that they will develop.

There are many descriptions of this process, most of which can be summed up as follows:

1. The size of the group is important. We have read of successful groups that range from 3 to 15; however, it is generally agreed that 4 to 8 is an optimum number for a brainstorming session.
2. Free expression is essential. That is what brainstorming is all about. Any evaluation of the exposed ideas is to be avoided. Nothing should be said to discourage a group member from speaking out.
3. The leader is a key figure, even though free expression is the hallmark. The leader sets the tone and tempo of the session and provides a stimulus when things begin to drag.
4. The members of the group should be equals. No one should feel any reason to impress or support any other member. If your supervisor also is a member, you must steer clear of concern for his or her feelings or support for his or her ideas.
5. Recorders are necessary. Everything that is said should be recorded, mechanically or manually. Evaluation comes later.

We have discussed a few techniques that are recommended to stimulate our thought processes. You may choose one of the free-wheeling techniques or perhaps a well-defined method. Regardless of your preferences, we think that you will be pleased and even surprised at the large list of ideas that you can develop in a short period of time.

2.7.3 The Chapter Example—Step 6

For this particular problem brainstorming was the method chosen by a large majority of the design teams. For reference the problem definition is repeated here.

Existing dorm beds → lofted beds

Figure 2.11

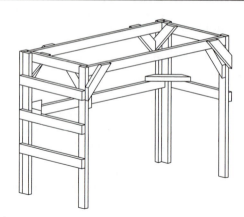

Alternative solution 1.

Since the problem definition is restricted only to lofted beds, the alternative solutions are for the loft system only and do not include a desk and shelves. Most
design teams started with a basic structure, then during the brainstorming they
came up with modifications that perhaps improved stability, were easier to assemble, used different material sizes, or seemed more functional. Five candidates for the best solution are presented here.

Figure 2.11 (alternative 1) shows a loft using two-by-four lumber for the
supporting structure. Stability is maintained by several supports, some of which
reduce the usable space under the loft. Nails are proposed for assembling.

An X-frame instead of vertical members are used in the solution presented
in Fig. 2.12 (alternative 2). This apparently reduces the amount of lumber but introduces instability. Note that the cross supports also serve as the ladder. Usable
space under the loft is very good. Screws or nails are to be used in assembly.

Figure 2.12

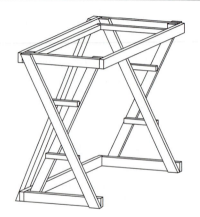

Alternative solution 2.

Figure 2.13

101

*Alternative
Solutions—Step 6*

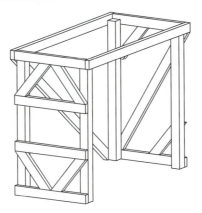

Alternative solution 3.

The solution shown in Fig. 2.13 (alternative 3) utilizes four-by-four posts as a base support. The ladder also serves as the cross support on one end. The design seems to be extremely stable but the material cost would be high. Screws are to be used in assembly.

Figure 2.14 (alternative 4) shows a simplistic design using four-by-four posts as a base support. This design also appears to be extremely stable. Screws are to be used in assembly.

The solution shown in Fig. 2.15 (alternative 5) is very similar to that of alternative 4 with less lumber used and an additional diagonal cross brace on one end. Screws are to be used in assembly.

The five alternative solutions are shown as concepts only. There are no dimensions specified, nor are any details of assembly. The lofts are sized overall to accommodate a standard bed size of 78 × 36 inches. The next step in the

Figure 2.14

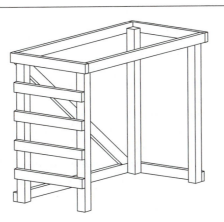

Alternative solution 4.

Figure 2.15

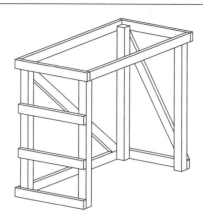

Alternative solution 5.

process is to analyze the five solutions, then to select the best one in view of the criteria.

2.8 Analysis—Step 7

At this point in the design process we have defined the problem, expanded our knowledge of the problem with a concentrated search for information, established constraints for a solution, selected criteria for comparing solutions, and generated alternative solutions. In order to find the best solution in light of available knowledge and criteria, we must analyze the alternative solutions to determine performance capability. Thus analysis becomes a pivotal point in the design process. Potential solutions which are not proved during the analysis phase may be discarded or, under certain conditions, retained with a redefinition of the problem and a change in constraints or criteria. Therefore one may need to repeat steps of the design process after completing the analysis.

Analysis involves the use of mathematical and engineering principles to determine the performance of a solution. Consider a system such as the cantilever beam in Fig. 2.16a, constrained by the laws of nature. When there is an input to the system (the applied load P), analysis will determine the performance of the system (beam)—that is, the deflection, stress buildup, and so on. Keep in mind that the objective of design is to determine the best solution (system) to a need; therefore knowledge of the performance of alternative solutions is necessary. We gain this knowledge through analysis.

The following example should be studied for the process rather than the content. This analysis involves the application of an engineering principle from the area of mechanics of materials. The goal is to obtain quantitative information for the decision step in the design process. Following this example we will describe the analysis used in the design of a lofted bed.

Example problem 2.1   Determine the deflection of the beam in Fig. 2.16a under the following conditions. Assume that the beam is structural steel.

$L = 4.0$ m (meters)

$h = 0.40$ m

$b = 0.20$ m

$P = 1.0 \times 10^5$ N (newtons)

Figure 2.16a

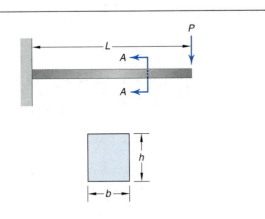

Solution   The deflection of the end of a cantilever beam for the configuration shown is given by

$$d = \frac{PL^3}{3\,EI} \text{ (constraint equation)}$$

where $d =$ deflection, m

$E =$ modulus of elasticity, a material constant, Pa

$\quad = 2.07(10^{11})$ Pa for structural steel

$I =$ moment of inertia, m^4

For a rectangular cross section

$$I = \frac{bh^3}{12}$$

$$= \frac{(0.2)(0.4)^3}{12}$$

$$= 1.067(10^{-3})\ m^4$$

Therefore

$$d = \frac{(10^5)\,(4)^3}{3(2.07)(10^{11})(1.067)(10^{-3})}$$

$$= 9.66\ (10^{-3})\ m$$

$$= 9.7\ mm$$

Figure 2.16b

	P, N	L, m	h,m	b,m	E, Pa	I, m^4	d,m
	1.00E+05	4	0.1	0.2	2.07E+11	1.67E-05	0.618357
	1.00E+05	4	0.2	0.2	2.07E+11	0.000133	0.077295
	1.00E+05	4	0.3	0.2	2.07E+11	0.00045	0.022902
	1.00E+05	4	0.4	0.2	2.07E+11	0.001067	0.009662
	1.00E+05	4	0.5	0.2	2.07E+11	0.002083	0.004947
	1.00E+05	4	0.6	0.2	2.07E+11	0.0036	0.002863
CANTILEVER BEAM DEFLECTION FOR RECTANGULAR SECTION							

This result would be forwarded to the designer for incorporation into the decision phase. The time required to produce an analysis is critical to the design process. If it takes longer to do an analysis than the schedule (Fig. 2.4) permits, then the results are somewhat meaningless. The engineer must exercise some judgment in selecting the method of analysis in order to assure results within the time limit. You can visualize the potential of computers in the analysis effort.

Many alternatives can be investigated in a brief amount of time. It is a simple task to change any of the parameters—b, h, P, L, or E—and to see the effect on the deflection immediately. Figure 2.16b shows a spreadsheet printout of cantilever beam deflections for a rectangular section for varying depths of the section while holding all other values constant. It would have been possible to produce plots of each parameter versus deflection by using computer graphics. Obviously the more possibilities one can investigate, the better the problem is understood and the better the design will be.

Analysis performed by engineers in most design projects is based on the laws of nature, the laws of economics, and common sense.

2.8.1   The Laws of Nature

You have already come into contact with many of the laws of nature, and you will no doubt be exposed to many more. At this point in your education you may have been exposed to the conservation principles: the conservation of mass, of energy, of momentum, and of charge. From chemistry you are familiar with the laws of Charles, Boyle, and Gay-Lussac. In mechanics of materials Hooke's law is a statement of the relationship between load and deformation. Newton's three principles serve as the basis of analysis of forces and the resulting motion and reactions.

Many methods exist to test the validity of an idea against the laws of nature. We might test the validity of an idea by constructing a mathematical model, for example. A good model will allow us to vary one parameter many times and to examine the behavior of the other parameters. We may very well determine the limits within which we can work. Other times we will find that our boundary conditions have been violated, and therefore the idea must be modified or discarded.

Results of an analysis of a mathematical model are frequently presented as graphs. Often the slopes of tangents to curves, points of intersection of curves, areas under or over or between curves, or other characteristics provide us with data that can be used directly in our designs.

Computer graphics enables a mathematical model to be displayed on a screen. As parameters are varied, the changes in the model and its performance also can be displayed quickly to the engineer.

The preparation of scale models of proposed designs often is a necessary step. These can be simple cardboard cutouts or they can involve the expenditure of great sums of money to test the models under simulated conditions that will predict how the real designs will perform under actual use. A prototype or pilot plant is sometimes justified because the cost of a failure is too great to chance. Such a decision usually comes only after other less expensive alternatives have been shown to be inadequate.

You probably have surmised that the more time and money that you allot to your model, the more reliable are the data that you receive. This fact is often distressing because we want and need good data but have to balance our needs against the available time and money.

2.8.2 The Laws of Economics

Section 2.8.1 introduced the idea that money and economics are part of engineering design and decision making. We live in a society that is based on economics and competition. It is probably true that many good ideas never get tried because they are deemed to be economically infeasible. Most of us have been aware of this condition in our daily lives. It started with our parents explaining why we could not have some item that we wanted because the item costs too much. Likewise, we will not include a very desirable component into our designs because the value gained will not return enough profit in relation to its cost.

Industry is continually looking for new products of all types. Some are desired because the current product is not competing well in the marketplace. Others are tried simply because it appears that people will buy them. How do manufacturers know that a new product will be popular? They seldom know with certainty. Statistics is an important consideration in market analysis. Most of you will find that probability and statistics are an integral part of your chosen engineering curriculum. The techniques of this area of mathematics allow us to make inferences about how large groups of people will react based on the reactions of a few. It is beyond our study at this time to discuss the techniques, but industry routinely employs such studies and invests millions of dollars based on their results.

2.8.3 Common Sense

We must never allow ourselves the luxury of failing to check our work. We also must judge the reasonableness of the result.

During the 1930s, the Depression years, a national magazine conducted a survey of voters and predicted a Republican victory. The magazine was wrong, and the public lost confidence in it to the point that the magazine went out of business. The editors had sampled the population by taking all the telephone books in the United States and, using a system of random numbers, selecting people to be called. They then applied good statistical analysis and made their prediction. Why did they miss so badly? It is a bit difficult today for us to imagine the Depression years, but the facts are that large percentages of voters did not have telephones, so this economic class of people was not included in the analysis. This group of people, however, did vote, and they voted largely Democratic. Nothing was wrong with the analytical method, only with the basic premise. The message is rather obvious: No matter how advanced our mathematical analysis, the results cannot be better than our basic assumptions. Similarly, we must always test our answers to see if they are reasonable.

2.8.4   The Chapter Example—Step 7

The analysis of the alternative loft systems was approached in several ways by student design teams. The efforts focused on two areas: general impressions of the proposed design in view of the criteria and specific considerations of the stability.

Following lengthly team discussions, general impressions of each alternative were compiled.

Alternative 1
- Use of two-by-fours decreases cost and weight.
- Requires significant cross bracing.
- Cross bracing interferes with usable space underneath loft bed.
- Time of assembly is high.

Alternative 2
- Uses minimum lumber.
- Has "different" look, may be appealing because of this.
- Lacks stability in longitudinal direction.

Alternative 3
- Has solid, stable construction with four-by-fours.
- Uses more lumber than other alternatives.

Alternative 4
- Has fairly easy assembly.
- Has "open" look.
- Uses minimum lumber.

Alternative 5
- Has fairly easy assembly.
- Has good stability.
- Has efficient use of lumber.

Specific consideration of the stability posed a problem for the student teams. There were no funds or time available to build and test prototypes of each of

Figure 2.17

107
Analysis—Step 7

A computer analysis of the deformations generated by a horizontal load applied to the loft bed.

the alternatives. The students did not yet have a background in engineering mechanics to perform structural analyses. Some teams decided to approach seniors in engineering for assistance. Seniors who had learned how to use structural analysis programs were able to construct a computer model for each alternative and to analyze the deformations to the structures when loads are applied. For the analyses the loft structures were assumed to be assembled with screws. This was a decision made by most teams during the establishment of the criteria. In considering the assembly requirements, the students deemed as important the difficulty of assembly translated into the time required for assembly. Assembly time is reduced in comparison to the use of bolts by using a power screwdriver and screws of appropriate length. Disassembly also would be easier if screws are used.

For structural analysis loads can be applied to a structure from any direction. Loads in the vertical direction would be determined by the weight of the person using the loft. Loads in the horizontal direction, applied to the end or side of the loft structure, would be estimated from someone pushing or leaning on the structure. Effects of combinations of loads also could be determined. Figure 2.17 shows the before and after shape (dashed outlines represent unloaded position) of alternative 5 after application of a horizontal load on the end. In this case the bottom was assumed to be fixed to the floor. A structural analysis software program called ANSYS was used. The deformation is very small and demonstrates high-level stability of that particular configuration.

We note here that beginning engineering students were not expected to conduct the structural analysis described earlier. The ingenuity of the student teams led them to find out more about how stability is determined. A strong characteristic of a successful engineer is the ability to seek out needed information from others that will help in the solution of the problem at hand.

As we review our own professional practices, we can honestly say that the most difficult times for us have not been when the analysis of a problem was difficult, but, rather, when it required a "tough" decision. We have known many engineers who are technically knowledgeable but who are unable to make a final decision. They may be happy to suggest several possible solutions and to outline the strong and weak points of each—indeed they may feel that their function is to do just that—but they let someone else decide which course is to be followed. The truth of the matter is that most engineering assignments require both functions: providing information and making decisions.

What makes reaching a decision so difficult? The answer is *trade-offs*. If we can be certain about anything in the future, it is that with your decisions the necessity to compromise will come. Review the criteria in Sec. 2.5.2 that were selected. If each alternative is evaluated against one criterion at a time, probably more than one alternative will surface as the best. Each time one criterion is optimized, another moves away from its optimum position. If the relationships are complicated, you may have to go through very complex processes to reach a decision. Certainly, no one idea will be better in all respects than all the others; hence you may have to choose a concept that you know is inferior to others in one or more of the decision criteria.

2.9.1 Organization for Decision

In order to decide among several alternatives, you need as much information as possible about each alternative. In design you need information in order to evaluate each alternative against each of the criteria. Analysis can provide the answers, as described in Sec. 2.7. If time and money are available, a prototype may be constructed and tested. In most cases judgment can be made with much less information. Computer models and engineering drawings are used to describe the form and function of the design.

Whatever information is available, it should fairly and accurately represent all the alternatives so that an equitable decision can be made.

2.9.2 Criteria in Decision

The objective of the entire design process is to choose the best solution for a problem within the time allowed. The steps that precede the decision phase are designed to give information that leads to the best decision. It should be quite obvious by now that poor research, a less than adequate list of alternatives, or inept analysis would reduce one's chances of selecting a good, much less than the best, solution. Decision making, like engineering itself, is both an art and a science. There have been significant changes during the past few decades that have changed decision making from being primarily an art to what it is at present, with probability, statistics, optimization, and utility theory all routinely used. Our purpose is not to explore these topics, but simply to note their influence and to consider for a moment our task of selecting the best of the proposed solutions to our problem. The term "optimization" is almost self-explanatory in that it emphasizes that what we seek is the best, or optimum, value in light of a criterion. As you study more mathematics, you will acquire more powerful tools through calculus and numerical methods for optimization.

In order to illustrate optimization, we will return to the beam problem illustrated in Fig. 2.16a and Example prob. 2.1. Our objective will be to determine the least mass of the beam for prescribed performance conditions. You will recall in our discussion of analysis that a system, the laws of nature and of economics, an input to the system, and an output are involved. Analysis gives the output if the system, laws, and input are known.

If we consider the inverse problem—that is, if we were looking for a system, given the laws, input, and output—we would be using *synthesis* rather than analysis. Synthesis is not as straightforward as analysis since it is possible to have more than one system that will perform as desired. But if we specify a criterion for selecting the best solution, then a unique solution is possible. The criterion used for selecting the best solution is often called a *payoff function*.

Example problem 2.2 (Refer to Fig. 2.16.) Determine the dimensions b, h for the least beam mass under the following conditions:

The deflection cannot exceed 40 mm.
The height h cannot be greater than three times the base b.

$$E = 2.0 \times 10^{11} \text{ Pa}$$

$$L = 4.0 \text{ m}$$

$$P = 1.0 \times 10^5 \text{ N}$$

If the beam has a constant cross section throughout, then the mass is a minimum when the cross-sectional area $A = bh$ is a minimum. Achieving minimum mass by finding the minimum area (payoff function) will provide the best (optimum) solution.

Solution The system we are after is the beam shape $b \times h$ within the conditions just specified; the law is the deflection equation from Example prob. 2.1; the inputs are L, P, and E, and the output is the range of permissible deflection. The deflection equation becomes

$$d = \frac{PL^3}{3EI}$$

$$0.04 = \frac{10^5 (4)^3}{3(2)(10^{11})I}$$

or

$$I = 2.667(10^{-4}) \text{ m}^4$$

Then

$$\frac{bh^3}{12} = 2.667(10^{-4})$$

Thus

$$b = \frac{3.2(10^{-3})}{h^3}$$

This equation is a relationship for the beam under the condition that the deflection is a constant 40 mm. The expression is plotted in Fig. 2.18a. Note that values to the right of the curve represent beam dimensions for which the deflection would be less than 40 mm. Those values to the left would cause the deflection to exceed 40 mm; thus that portion of the *design space* for b and h is invalid.

Next we demonstrate the effect of the required relationship between b and h by plotting the line $b = h/3$, as shown in Fig. 2.18a. Points above this line represent valid geometric configurations; those below do not.

Now we have a better picture of the design space, or solution region, for our problem. A point $h = 0.3$ m, $b = 0.3$ m, represents a satisfactory solution since it falls within all conditions *except* possibly minimum mass. Many designs stop at this point when a nominal solution has been found. These are the designs that may not survive in the marketplace because they are not optimum. In fact, to get a nominal solution, we could have guessed values for b and h and very quickly had an answer without going to all the effort we have up to this point.

We know the region in which the best solution (minimum area) lies. We again take advantage of the capability of the spreadsheet and have it search for the solution we want. Figure 2.18b shows a table of values for h and the two constraints on the solution, namely, a deflection of 40 mm or less and that h cannot be greater than 3 times b. We set up the spreadsheet to iterate on the payoff function, $A = bh$, within the constraints, and find a minimum. The solution in this case is the intersection of the two curves in Fig. 2.18a. Often we find that optimum solutions lie on the boundary of the design solution space. We also note that the function $A = bh$ does not have a minimum since it is asymptotic to the coordinate axes. However, in most design applications the payoff function will be of a complex nature and will not have obvious shape characteristics.

You also will note that the spreadsheet analysis can be quickly extended to analyze cantilever beam designs for any range of loading, sizes, and materials simply by replacing the values in the governing equations. This capability will be valuable in your upcoming engineering coursework.

2.9.3 The Chapter Example—Step 8

The decision matrix was the method chosen for determining the best loft structure. The matrix, shown in Fig. 2.19, enables a fair comparison of all the alternatives by the design team and is relatively straightforward to develop. It is a structured procedure that enables the individual team members to have equal input to the decision process. The first two columns list the criteria and their respective weights in percent. The matrix is filled in by going across and evaluating each alternative against a single criterion. For example, alternative 1 received a 4 (poor) for ease of assembly. When the rating is multiplied by the weight, a point value of 140 is obtained. Similarly, alternative 2 received a 5

Figure 2.18a

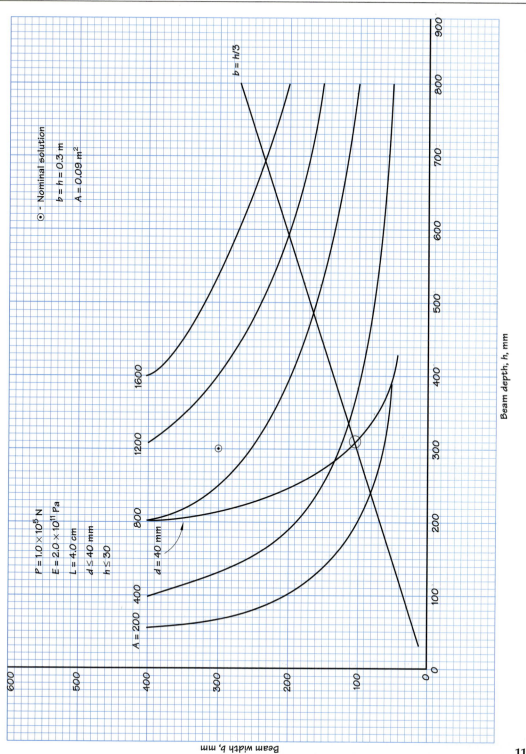

Beam depth, *h*, mm

Beam width *b*, mm

$\odot$ - Nominal solution
$b = h = 0.3$ m
$A = 0.09$ m²

$b = h/3$

$A = 1600$
$A = 1200$
$A = 800$
$d = 40$ mm
$d \le 40$ mm
$A = 400$
$A = 200$

$P = 1.0 \times 10^5$ N
$E = 2.0 \times 10^{11}$ Pa
$L = 4.0$ cm
$d \le 40$ mm
$h \le 30$

Figure 2.18b

	h, m		0.0032/h^3		b=h/3	
	0.15		0.948148		0.05	
	0.2		0.4		0.066667	
	0.25		0.2048		0.083333	
	0.3		0.118519		0.1	
	0.35		0.074636		0.116667	
	0.4		0.05		0.133333	
	Minimum area = 0.03266 m^2 when h = 0.313017 m and b = 0.104339 m					

Figure 2.19

Criteria	Weight, W%	Alternative Solutions						
		1	2	3	4	5	6	7
Ease of assembly	35	4 / 140	5 / 175	6 / 210	8 / 280	8 / 280		
Functionality	25	5 / 125	8 / 200	8 / 200	8 / 200	8 / 200		
Cost	25	6 / 150	6 / 150	5 / 125	7 / 175	7 / 175		
Stability	15	7 / 105	3 / 45	9 / 135	9 / 135	10 / 150		
Total	100	520	570	670	790	805		

Rating scale R

Excellent	9–10
Good	7–8
Fair	5–6
Poor	3–4
Unsatisfactory	0–2

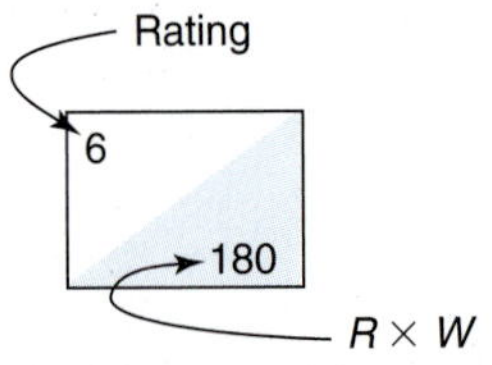

Rating

6 / 180

$R \times W$

Each concept was rated by the team on a scale of 0 to 10 for each criterion. The rating was multiplied by the criterion weight and then summed. Alternative 5 was chosen as the optimum.

rating for the same criterion for a point total of 175. After each alternative was evaluated for ease of assembly, the alternatives were then evaluated on functionality, and so on until the matrix is filled. Total points for each alternative are determined, and the one with the highest point value would be considered the best solution

Alternatives 4 and 5 had the same rating except for a slight difference in stability. Because the totals are so close, and the ratings are qualitative in nature, either solution would be acceptable. It is obvious that alternatives 4 and 5 were far ahead of the others. In general most design teams will choose the alternative with the highest point total, but they do so only after redoing the decision matrix for just those two alternatives and reviewing the search and criteria steps carefully. The additional columns shown in Fig. 2.19 may be used to generate new alternatives which are a combination of two or more of the original alternatives. Often a new combination will generate a higher rating when the best features of two or more alternatives are combined.

2.10 Specification—Step 9

Once the decision has been made, many engineers would like to believe the work is completed. The suspense and uncertainty of the solution are over, but much work still lies ahead. Even if the new idea is not a breakthrough in technology but simply an improvement in existing technology, it must be clearly defined to others. Many very creative people are ill-equipped to convey to others just exactly what solution they are proposing. It is not the time to use vague generalities about the general scope and to approximate size and shape of the chosen concept. One must be extremely specific about all the details regardless of the apparent minor role that each may play in the finished product.

2.10.1 Graphical Specifications

You will probably have many occasions in which to work closely with technicians and drafters as they prepare the countless databases that are essential to the manufacture of your design. You will not be able to do your job properly if you cannot sketch well enough to portray your idea or to interpret databases well enough to know whether the plans that you must approve will actually result in your idea being constructed as you desire.

A lathe operator in the shop, an electronics technician, a contractor, or someone else must produce your design. How is the person to know what the finished product is to look like, what materials are to be used, what thicknesses are required, how it is to work, what clearances and tolerances are demanded, how it is to be assembled, how it is to be taken apart for maintenance and repair, what fasteners are to be used, and so on?

Typical databases that are normally required include:

1. A sufficient number of databases describing the size and shape of each part.
2. Layouts to delineate clearances and operational characteristics.
3. Assembly and subassemblies to clarify the relationship of parts.

4. Written notes, standards, specifications, and so on, concerning quality and tolerances.

5. A complete bill of materials.

Included with the drawings are written specifications, although certain classes of engineering work refer simply to documented standard specifications. For example, most cities have adopted one of several national building codes, so all structures constructed in those cities must conform to the code. It is quite common for engineers and architects to refer simply to the building code as part of the written specifications and to write detailed specifications for only those items that are not covered in the code. This procedure saves time and money for all by providing uniformity in bidding procedures. Many groups have produced standards that are widely recognized. For instance, there are standards for welds and fasteners for the obvious reasons—ease of specification and economy of manufacture. Moreover, there are such standards for each discipline of engineering.

Thus the engineer must be able to utilize the written, spoken, and graphical languages in order to develop and interpret specifications. There can be no ambiguities and no room for error because mistakes in judgment or interpretation cost money. In your early months as a new graduate working in the industry you will spend a significant amount of time studying and learning the standard for specifications in your field. You also may have the opportunity to write new or upgraded standards.

2.10.2 The Chapter Example—Step 9

We will present a representative sample of the specifications for the loft bed. Figures 2.20a and b are detailed drawings of the two diagonal cross braces. From these drawings the two-by-fours can be cut to length and then the angle cuts can be made. Figure 2.21 shows the overall dimensions of the loft from which the required lengths of the component parts can be determined. Figure 2.22 is the bill of material for the loft and is based on the available standard lengths of lumber. There will be some waste unless the lumber yard is

Figure 2.20a

Detailed drawing of back diagonal brace.

Figure 2.20b

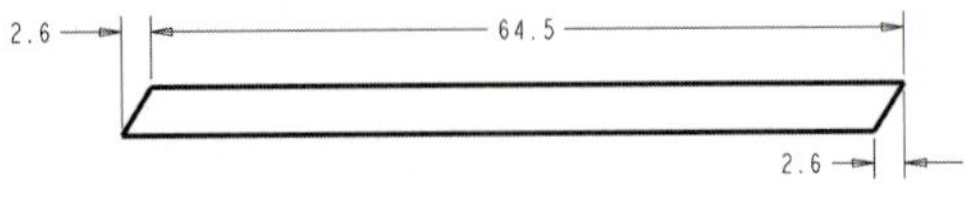

Detailed drawing of end diagonal brace.

Figure 2.21

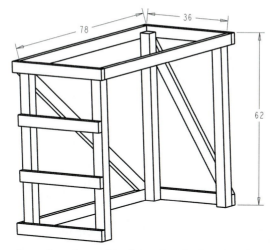

A pictorial with key dimensions is used to determine the lengths of the components.

Figure 2.22

Bill of Material

Item	Amount Required	Cost*, $
2" × 4" × 16'	3	18.00
2" × 4" × 8'	1	2.89
4" × 4" × 8'	4	32.00
3" deck screws	1 lb.	2.50
		55.39

*Applicable taxes are not included.

Bill of material for loft bed. Cost is included.

asked to cut the lumber prior to purchase. This would add some cost to the loft. Note that the design teams have not included any tools in the bill of material, assuming that students could take advantage of the tool checkout at the dormitories.

2.11 Communication—Step 10

During the 1960s the word "communication" seemed to take on a very high priority at conferences and in professional journals. The need for conveying information and ideas had not changed, but there was an awareness of too much incomplete and inaccurate rendering of information. At most of the professional conferences that the authors have attended, one or more papers either discussed the need for engineers to develop greater skills in communication or demonstrated a technique for improving the skills. Students at most universities are required to complete freshman English courses and, in some colleges, a technical writing course in their junior or senior year. However, many professors and employers feel that not enough emphasis is being placed on the application of communication skills.

For our purposes here, though, we will discuss only the salient points involved in design step 10.

2.11.1 Selling the Design

It is certainly the responsibility of any profession to inform people of findings and developments. Engineering is no exception in this regard. Our emphasis here will be on a second type of communication, however: selling, explaining, and persuading.

Selling takes place all the way through the design process. Individuals who are the most skillful at it will see many of their ideas develop into realities. Those who are not so good at selling will no doubt become frustrated with their supervisors for not exploring in more depth what they feel is a perfectly good idea.

If you are working as a design engineer for an industry, you cannot simply decide on your own that you will try to improve the product line. Industry is anxious for its engineers to initiate ideas but will not necessarily approve all of them. As an engineer with a company, you must convince those who decide what assignments you get that the idea is worth the time and money required to develop it. Later, after the design has developed to the point where it can be produced and tested, you must persuade management again to place it into production.

A natural reaction is to feel that your design has so many clear advantages that your selling it to management should not be necessary. This may be the case, but in actual situations things seldom work so smoothly and simply. You will be selling or persuading or convincing others almost daily in a variety of ways. Among the many forms of communication are written and oral reports.

2.11.2 The Written Report

The types of reports that you will write as an engineer will be varied, so a precise outline that will serve for all the reports cannot be supplied. The two major types of reports are those used by individuals within the organization and those used primarily by clients or customers. Many times the in-house reports follow a strict form prescribed by the organization, whereas those intended for the client are usually designed for the particular situation. The nature of the project and the client usually determine the degree of formality employed in the report. Clients often state that they wish to use the report in some particular manner, which may direct you to the style of report to use. For instance, if you are a consultant for a city and have studied the needs for expansion of a power plant, the report may be very technical, brief, and full of equations, computations, and so on, if it is intended for the use of the city's engineer and public utilities director. However, if the report is to be presented to citizens in an effort to convince them to vote for a bond issue to finance the expansion, the report will take a different flavor.

Reports generally will have the following divisions or sections:

1. Appropriate cover page
2. Abstract
3. Table of contents
4. Body
5. Conclusions and recommendations
6. Appendixes

Abstract. A brief paragraph indicating the purpose and results of the effort being reported, the abstract is used primarily for archiving so that others can decide quickly if they want to obtain the complete report.

Body. This is the principal section of the report. It begins with an introduction to activities, including problem identification, background material, and the plan for attacking and solving the problem. If tests were conducted, research completed, and surveys undertaken, the results are recounted and their significance is underscored. In essence, the body of the report is the description of the individual or team effort on a project.

Conclusions and Recommendations. This section gives the reason for conducting the study and explains the purpose of the report. Herein explain what you now believe to be true as a resut of the work discussed in the body and what you recommend should be done about it. You must lay the groundwork earlier in the report, and at this point you must sell your idea. If you have done the job carefully and fully, you may make a sale; however, do not be discouraged if you do not. There will be other days and other projects.

Appendixes. Appendixes can be used to avoid interrupting your descriptions so that it can flow more smoothly. Those who do not want to know everything about your study can read it without digression. What is in the appendix completes the story by showing all that was done. But it should not contain information that is essential to one's understanding of the report.

It should be emphasized that all reports do not follow a specific format. For example, lengthy reports should have a summary section placed near the beginning. This one- or two-page section should include a brief statement of the problem, the proposed solution, the anticipated costs, and the benefits. The summary is for the use of higher-level management who in general do not have the time to read your entire report.

In many instances your instructor or supervisor will have specific requirements for a report. Each report is designed to accomplish a specific goal.

Student reports often must follow an instructor's directions regarding form and topics that must be included. If you are asked to write a report in an introductory design course, refer to Fig. 2.1 to make sure that all the steps in the design process have been successfully completed. You also might study the bias of your instructor and make sure that you have done especially the steps that he or she considers most important. This may sound as though pleasing the instructor and getting a good grade is all that is important. But perhaps in this respect the academic situation is something like that in industry or private practice: Your report must take into account the audience—its biases and its expectations, whether a professor in the classroom or a supervisor in the business world.

2.11.3 The Oral Presentation

The objective of the oral presentation is the same as that of the written report—to furnish information and to convince the listener. However, the methods and techniques are quite different. The written report is designed to be glanced at, read, and then studied. The oral presentation is a one-shot deal that must be done quickly, so it must be simple. There is no time to go into detail, to show complicated graphs and tables of data or many of the things that are given in a written report. What can you do to make a good presentation?

First, you must be prepared. No audience listens to people who have not bothered to prepare themselves. So you should rehearse with a timer, a mirror, and a tape recorder.

Stand in such a way so that you do not detract from what you are saying or showing.

Look at your audience and maintain eye contact. You will be receiving cues from those who are listening, so be prepared to react to these cues.

Project your voice by consciously speaking to the back row. The audience quickly loses interest if it has to struggle to hear.

Speak clearly. We all have problems with our voices—they are either too high, too low, or too accented and certain words or sounds are hard for us, but always be concerned for the listener.

Preparation obviously includes being thoroughly familiar with the material. It should also include determining the nature, size, and technical competence of the audience. You must know how much time will be allotted to your presentation and what else, if anything, is to be presented before or after your speech. It is essential that you know what the room is like because the physical conditions of the room—its size, lighting, acoustics, and seating arrangements—may very well control your use of slides, transparencies, video, and microphones.

The quality of your graphic displays often can influence the opinion of your audience. Again, consider to whom you are speaking carefully as you choose which and how many displays to use. Be certain that they can be read and understood or do not use them at all. Do not clutter your displays with so many details that the message is obscured. Do not try to make a single visual aid accomplish too many tasks: It is good to change the center of emphasis. By all means, test your visual aids before the meeting and never apologize for their quality. (If they are not good, do not use them.) Figure 2.23 shows an exploded pictorial of a small battery-powered grinding tool and provides a good overview of the tool's components.

Figure 2.24 narrows the focus on a subassembly of the tool. The quality of your visual aids can influence many people for you or against you before they hear all you have to say.

Have a good finish. Save something important for the last and make sure everyone knows when the end has come. Be sure not to end with "Well, I guess

Figure 2.23

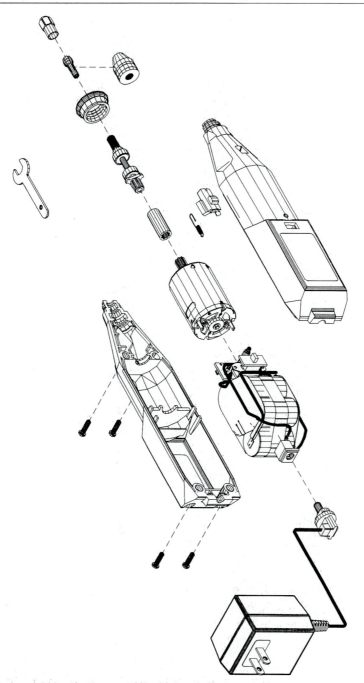

Exploded pictorial shows all parts of the high-speed rotary tool.

Figure 2.24

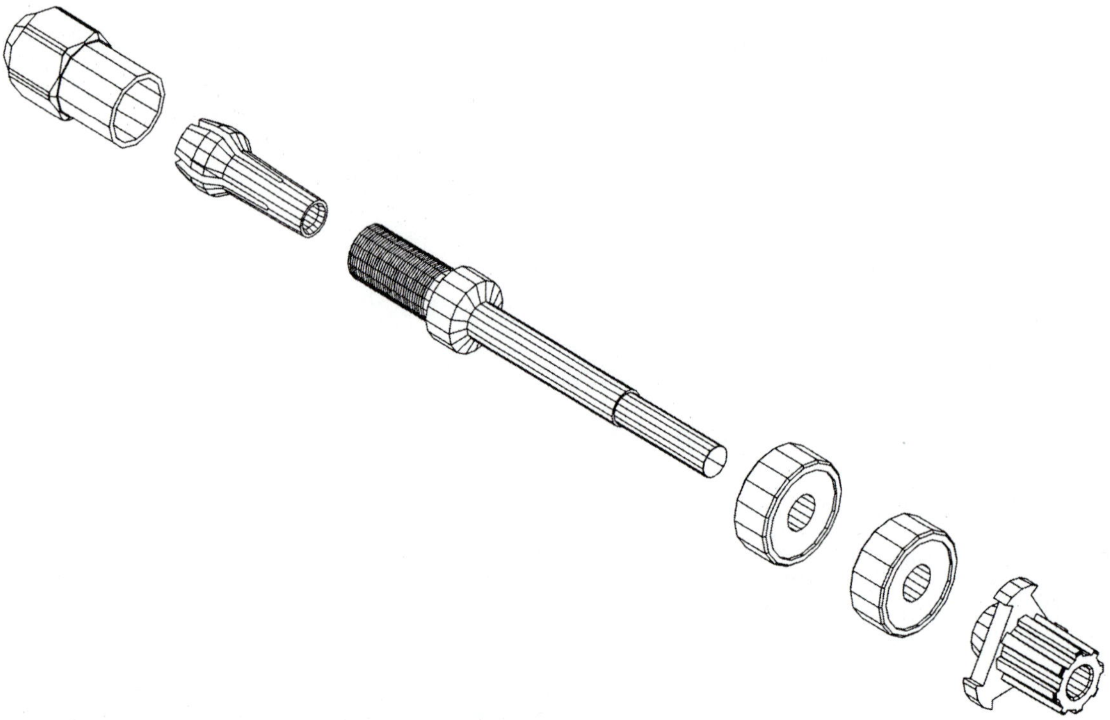

Subassembly of rotary tool showing bearings, shaft, and collet.

that's about all I have to say." You have much more to say but just do not have the time to say it.

2.11.4 The Chapter Example—Step 10

The written and oral reports presented by the student design teams are significant parts of their design experience. Reports are regulated somewhat by their professors in much the same way that reports are in industry. Students are told who would be reading the report and who would judge the oral presentation; they are given copies of the written report grading sheet and the oral presentation judging card, and they are evaluated on their execution of the design process as well as on the written and oral presentations.

Key Terms and Concepts

We have described the engineering design process with a series of 10 steps. The application of the process to an engineering design process, although structured, is an iterative process with flexibility to make necessary adjustments as the design progresses. The emphasis in this chapter is on conceptual design. You should realize that many of the details of specification and communication of the loft design are not included.

At this stage of your engineering education it is important that you un-
dergo the experience of applying the design process to a need with which you can identify based on your personal experiences. As you approach the bac- calaureate degree you will have acquired the technical capability to conduct the necessary analyses and to make the appropriate technical decisions required for complex products, systems, and processes.

The following are some of the terms and concepts you should recognize and understand.

Engineering design	**Brainstorming**
Design process	**Solution space**
Bloom's Taxonomy on Learning	**Analysis**
Customer satisfaction	**Synthesis**
Constraint	**Payoff function**
Criteria	**Decision matrix**
Reverse engineering	**Specification**
Alternative solutions	**Communication**
Checkoff lists	

Problems

Problems 2.1 and 2.2 involve synthesis, which is similar to that illustrated in Example prob. 2.2.

2.1 For beam configuration of Fig. 2.16 a, determine the dimensions b, h for the least mass under the following conditions:

The deflection cannot exceed 50 mm.
The height h cannot be greater than b.

$$E = 2.0 \times 10^{11} \text{ Pa}$$

$$L = 6.0 \text{ m}$$

$$P = 1.0 \times 10^5 \text{ N}$$

Produce a brief report containing a computer plot that is similar to Fig. 2.18a and a discussion of the design space and how the solution was found.

2.2 A company transfers packages from point to point across the country. The limit on package size is that the girth plus the longest dimension (measured on the package) cannot exceed 130 inches. Consider two kinds of packages, a rectangu- lar prism shape with square ends and a cylindrical shape where the cylinder height is greater than the diameter (girth equals the circumference). Determine for each shape the largest package volume that can be shipped and the package dimensions at the maximum volume. A *suggested procedure* follows:
(a) Write the constraint equation (130 inch limit) and the payoff function (vol- ume). Eliminate one of the unknowns in the payoff function by substituting the constraint equation.
(b) Plot the payoff function against the remaining variable.
(c) Determine the optimum values.
(d) Prepare a report of your findings.
2.3 Investigate current designs for one or more common items. If you do not have the items in your possession, purchase them or borrow from friends. Conduct the following "reverse engineering" procedures on each of the items:
(a) Write down the need that the design satisfies.
(b) Disassemble the item and list all the parts by name.
(c) Write down the function of each of the parts in the item.

(d) Reassemble the item.

(e) Write down answers to the following questions:

- Does the item satisfactorily solve the need you stated in part *(a)*?
- What are the strengths of the design?
- What are the weaknesses of the design?
- Can this design be easily modified to solve other needs? If so, what needs and what modifications should be made?
- What other designs can solve the stated need?

The items for your study are the following:

- Mechanical pencil
- Safety razors from three vendors; include one disposable razor
- Flashlight
- Battery-powered slide viewer
- Battery-powered fabric shaver

2.4 For the chapter example, analyze in detail one or more of the alternatives that did not get selected as the best. Produce design modifications that may possibly enable these alternatives to receive higher ratings in the decision matrix. Develop a brief report and oral presentation of your results.

2.5 Beginning with alternative 5, the loft bed that was selected as the best solution, complete the loft system by designing a desk and shelves. Follow the design process and select the best overall system from a series of alternatives. Develop a report and oral presentation of your results.

2.6 The following list of potential projects can be approached in the manner used by the student design team featured in this chapter. Develop a report and oral presentation as directed by your instructor.

- Headlights that follow the wheels' direction
- A protective "garage" that can be stored in the car's trunk
- A device to prevent the theft of helmets left on motorcycles
- A conversion kit for winter operation of motorcycles
- An improved rack for carrying packages or books on a motorcycle or bicycle
- A child's seat for a motorcycle or bicycle
- A tray for eating, writing, and playing games in the back seat of a car
- A system for improving traction on ice without studs or chains
- An inexpensive built-in jack for raising a car
- An auto engine warmer
- A better way of informing motorists of speed limits, road conditions, hazards, and so on
- Theft- and vibration-proof wheel covers
- A better way to check the engine oil level
- A device to permit easier draining of the oil pan by weekend mechanics
- A heated steering wheel for cold weather
- A less expensive replacement for auto air-cleaner elements
- An overdrive system for a trail bike
- A sun shield for an automobile
- A well-engineered, efficient automobile instrument panel
- An SOS sign for cars stalled on freeways
- A remote car-starting system for warm-up
- A car-door positioner for windy days
- A bicycle trailer
- Automatic rate-sensitive windshield wipers

- A storage system for a cell phone in a car (including charger)
- A corn detasseler
- An improved wall outlet
- A beverage holder for a card table
- A car wash for pickups
- A better rural mailbox
- A home safe
- An improved automobile traffic pattern on campus
- An alert for drowsy or sleeping drivers
- An improved bicycle for recreation or racing
- Improved bicycle brakes
- A transit system for campus
- A pleasure boat with retractable trailer wheels
- Improved pedestrian crossings at busy intersections
- A transportation system within large airports
- An improved baggage-handling system at airports
- Improved parking facilities in and around campus
- A simple but effective device to assist in cleaning clogged drains
- A device to attach to a paint can for pouring
- An improved soap dispenser
- A better method of locking weights to a barbell shaft
- A shoestring fastener to replace the knot
- An automatic moisture-sensitive lawn waterer
- A better harness for Seeing Eye dogs
- A better jar opener
- A system or device to improve efficiency of limited closet space
- A shoe transporter and storer
- A pen and pencil holder for college students
- An acceptable rack for mounting electric fans in dormitory windows
- A device to pit fruit without damage
- A riot-quelling device to subdue participants without injury
- An automatic device for selectively admitting and releasing an auxiliary door for pets
- A device to permit a person to open a door when loaded with packages
- A more efficient toothpaste tube
- A fingernail catcher for fingernail clippings
- A more effective alarm clock for reluctant students
- An alarm clock with a display to show it has been set to go off
- A device to help a parent monitor small children's presence and activity in and around the house
- A chair that can rotate, swivel, rock, or stay stationary
- A simple pocket alarm that is difficult to shut off, used for discouraging muggers
- An improved storage system for luggage, books, and so on in dormitories
- A lampshade designed to permit one to study while his or her roommate is asleep
- A device that would permit blind people to vote in an otherwise conventional voting booth
- A one-cup coffeemaker
- A solar greenhouse
- A quick-connect garden-hose coupling
- A device for recycling household water
- A silent wake-up alarm
- Home aids for the blind (or deaf)

- A safer, more efficient, and quieter air mover for room use
- A can crusher
- A rain-sensitive house window
- A better grass catcher for a riding lawn mower
- A winch for hunters of large game
- Gages for water, transmission fluid, and so on in autos
- A built-in auto refrigerator
- A better camp cooler
- A dormitory cooler
- A device for raising and lowering TV racks in the classroom
- An impact hammer adapter for electric drills
- An improved method of detecting and controlling the level position of the bucket on a bucket loader
- Shields to prevent corn spillage where the drag line dumps into the sheller elevator (angle varies)
- An automatic tractor-trailer-hitch aligning device
- A jack designed expressly for motorcycle use (special problems involved)
- A motorbike using available (junk) materials
- Improved road signs for speed limits, curves, deer crossings, and so on
- More effective windshield wipers
- A windshield deicer
- Shock-absorbing bumpers for minor accidents
- A home fire-alarm device
- A means of evacuating buildings in case of fire
- Automatic light switches for rooms
- A carbon monoxide detector
- An indicator to report the need for an oil change
- A collector for dust (smoke) particles from stacks
- A means of disposing of or recycling soft-drink containers
- A way to stop dust storms, resultant soil loss, and air entrainment
- An attractive system for handling trash on campus
- A self-decaying disposable container
- A device for dealing with oil slicks
- A means of preventing heat loss from greenhouses
- A way of creating energy from waste
- A bookshelf with horizontally and vertically adjustable shelves and dividers
- An egg container (light, strong, compact) for camping and canoeing
- Ramps or other facilities for handicapped students
- A multifunctional (suitcase/chair/bookshelf, etc.) packing device for students
- An adapter to provide tilt and elevation control on existing graphics tables
- A compact and inexpensive camp stove for backwoods hiking
- A road trailer operable from inside the car
- A hood lock for cars to prevent vandalism
- A system to prevent car thefts
- A keyless lock

CHAPTER 3

Engineering Solutions

3.1 Introduction

The practice of engineering involves the application of accumulated knowledge and experience to a wide variety of technical situations. Two areas, in particular, that are fundamental to all of engineering are design and problem solving. The professional engineer is expected intelligently and efficiently to approach, analyze, and solve a range of technical problems. These problems can vary from single solution, reasonably simple problems to extremely complex, open-ended problems that require a multidisciplinary team of engineers.

Problem solving is a combination of experience, knowledge, process, and art. Most engineers through either training or experience solve many problems by a process. The design process, for example, is a series of logical steps that when followed produce an optimal solution given time and resources as two constraints. The total quality (TQ) method is another example of a process. This concept suggests a series of steps leading to desired results while exceeding customer expectations.

This chapter provides a basic guide to problem analysis, organization, and presentation. Early in your education you must develop an ability to solve and present simple or complex problems in an orderly, logical, and systematic way.

3.2 Problem Analysis

A distinguishing characteristic of a qualified engineer is the ability to solve technical problems. Mastery of problem solving involves a combination of art and science. By *science* we mean a knowledge of the principles of mathematics, chemistry, physics, mechanics, and other technical subjects that must be learned so that they can be applied correctly. By *art* we mean the proper judgment, experience, common sense, and know-how that must be used to reduce a real-life problem to such a form that science can be applied to its solution. To know when and how rigorously science should be applied and whether the resulting answer reasonably satisfies the original problem is an art.

Much of the science of successful problem solving comes from formal education in school or from continuing education after graduation. But most of the art of problem solving cannot be learned in a formal course; rather, it is a result of experience and common sense. Its application can be more effective, however, if problem solving is approached in a logical and organized method— that is, if it follows a process.

To clarify the distinction, let us suppose that a manufacturing engineer and a logistics specialist working for a large electronics company are given the task of recommending whether the introduction of a new computer that will focus on the computer-aided-design (CAD) market can be profitably produced. At the time this task is assigned the competitive selling price has already been estimated by the marketing division. Also, the design group has developed working models of the computer with specifications of all components, which means that the approximate cost of these components is known. The question of profit thus rests on the costs of assembly and distribution. The theory of engineering economy (the science portion of problem solving) is well known and applicable to the cost factors and time frame involved. Once the production and distribution methods have been established, these costs can be computed using standard techniques. Selection of production and distribution methods (the art portion of problem solving) depends largely on the experience of the engineer and logistics specialist. Knowing what will or will not work in each part of these processes is a must in the cost estimate; however; these data cannot be found in handbooks, but, rather, they are found in the minds of the logistics specialist and the engineer. It is an art originating from experience, common sense, and good judgment.

Before the solution to any problem is undertaken, whether by a student or a practicing professional engineer, a number of important ideas must be considered. Consider the following questions: How important is the answer to a given problem? Would a rough, preliminary estimate be satisfactory, or is a high degree of accuracy demanded? How much time do you have and what resources are at your disposal? In an actual situation your answers may depend on the amount of data available or the amount that must be collected, the sophistication of equipment that must be used, the accuracy of the data, the number of people available to assist, and many other factors. Most complex problems require some level of computer support such as a spreadsheet or a math analysis program. What about the theory you intend to use? Is it state of the art? Is it valid for this particular application? Do you currently understand the theory, or must time be allocated for review and learning? Can you make assumptions that simplify without sacrificing needed accuracy? Are other assumptions valid and applicable?

The art of problem solving is a skill developed with practice. It is the ability to arrive at a proper balance between the time and resources expended on a problem and the accuracy and validity obtained in the solution. When you can optimize time and resources versus reliability, then problem-solving skills will serve you well.

3.3 The Engineering Method

The *engineering method* is an example of process. It consists of six basic steps:

1. *Recognize and understand the problem.*   Perhaps the most difficult part of problem solving is developing the ability to recognize and define the problem precisely. This is true at the beginning of the design process and when applying the engineering method to a subpart of the overall problem. Many academic problems that you will be asked to solve have this step completed by the instructor. For example, if your instructor asks you to solve a quadratic algebraic equation and provides all the coefficients, the problem has been completely defined before it is given to you and little doubt remains about what is the problem.

If the problem is not well defined, considerable effort must be expended at the beginning in studying the problem, eliminating the things that are unimportant, and focusing on the root problem. Effort at this step pays great dividends by eliminating or reducing false trials, thereby shortening the time taken to complete later steps.

2. *Accumulate data and verify accuracy.* All pertinent physical facts, such as sizes, temperatures, voltages, currents, costs, concentrations, weights, times, and so on, must be ascertained. Some problems require that steps 1 and 2 be done simultaneously. In others, step 1 might automatically produce some of the physical facts. Do not mix or confuse these details with data that are suspect or only assumed to be accurate. Deal only with items that can be verified. Sometimes it will pay actually to verify data that you believe to be factual but actually may be in error.

3. *Select the appropriate theory or principle.* Select appropriate theories or scientific principles that apply to the solution of the problem; understand and identify limitations or constraints that apply to the selected theory.

4. *Make necessary assumptions.* Perfect solutions to real problems do not exist. Simplifications need to be made if they are to be solved. Certain assumptions can be made that do not significantly affect the accuracy of the solution, yet other assumptions may result in a large reduction in accuracy.

Although the selection of a theory or principle is stated in the engineering method as preceding the introduction of simplifying assumptions, there are cases where the order of these two steps should be reversed. For example, if you were solving a material balance problem, you often need to assume that the process is steady, uniform, and without chemical reactions so that the applicable theory can be simplified.

5. *Solve the problem.* If steps 3 and 4 have resulted in a mathematical equation (model), it is normally solved by an application of mathematical theory, although a trial-and-error solution which employs the use of a computer or perhaps some form of graphical solution also may be applicable. The results normally will be in numerical form with appropriate units.

6. *Verify and check results.* In engineering practice, the work is not finished merely because a solution has been obtained. It must be checked to ensure that it is mathematically correct and that units have been properly specified. Correctness can be verified by reworking the problem by using a different technique or by performing the calculations in a different order to be certain that the numbers agree in both trials. The units need to be examined to ensure that all equations are dimensionally correct. And finally, the answer must be examined to see if it makes sense. An experienced engineer will generally have a good idea of the order of magnitude to expect.

If the answer doesn't seem reasonable, there is probably an error in the mathematics, in the assumptions, or perhaps in the theory used. Judgment is critical. For example, suppose that you are asked to compute the monthly payment required to repay a car loan of $5 000 over a 3-year period at an annual interest rate of 12 percent. Upon solving this problem, you arrived at an answer of $11 000 per month. Even if you are inexperienced in engineering economy, you know that this answer is not reasonable, so you should reexamine your theory and computations. Examination and evaluation of the reasonableness of an answer is a habit that you should strive to acquire. Your instructor and employer alike will find it unacceptable to be given results which you have indicated to be correct but are obviously incorrect by a significant percentage.

3.4 Problem Presentation

The engineering method of problem solving as presented in the previous section is an adaptation of the well-known *scientific problem-solving method*. It is a time-tested approach to problem solving that should become an everyday part of the engineer's thought process. Engineers should follow this logical approach to the solution of any problem while at the same time learning to translate the information accumulated into a well-documented problem solution.

The following steps parallel the engineering method and provide reasonable documentation of the solution. If these steps are properly executed during the solution of problems in this text and all other courses, it is our belief that you will gradually develop an ability to solve and properly document a wide range of complex problems.

1. *Problem statement.* State as concisely as possible the problem to be solved. The statement should be a summary of the given information, but it must contain all essential material. Clearly state what is to be determined. For example, find the temperature (K) and pressure (Pa) at the nozzle exit.
2. *Diagram.* Prepare a diagram (sketch or computer output) with all pertinent dimensions, flow rates, currents, voltages, weights, and so on. A diagram is a very efficient method of showing given and needed information. It also is an appropriate way of illustrating the physical setup, which may be difficult to describe adequately in words. Data that cannot be placed in a diagram should be listed separately.
3. *Theory.* The theory used should be presented. In some cases a properly referenced equation with completely defined variables is sufficient. At other times an extensive theoretical derivation may be necessary because the appropriate theory has to be derived, developed, or modified.
4. *Assumptions.* Explicitly list in complete detail any and all pertinent assumptions that have been made to realize your solution to the problem. This step is vitally important for the reader's understanding of the solution and its limitations. Steps 3 and 4 might be reversed or integrated in some problems.
5. *Solution steps.* Show completely all steps taken in obtaining the solution. This is particularly important in an academic situation because your reader, the instructor, must have the means of judging your understanding of the solution technique. Steps completed but not shown make it difficult for evaluation of your work and therefore difficult to provide constructive guidance.
6. *Identify results and verify accuracy.* Clearly identify (double underline) the final answer. *Assign proper units.* An answer without units (when it should have units) is meaningless. Remember, this final step of the engineering method requires that the answer be examined to determine if it is realistic, so check solution accuracy and, if possible, verify the results.

Once the problem has been solved and checked, it is necessary to present the solution according to some standard. The standard will vary from school to school and industry to industry.

On most occasions your solution will be presented to other individuals who are technically trained, but you should remember that many times these individuals do not have an intimate knowledge of the problem. However, on other occasions you will be presenting technical information to persons with nontechnical backgrounds. This may require methods that are different from those used to communicate with other engineers; thus it is always important to understand who will be reviewing the material so that the information can be clearly presented.

One characteristic of engineers is their ability to present information with great clarity in a neat, careful manner. In short, the information must be communicated accurately to the reader. (Discussion of drawings or simple sketches will not be included in this chapter, although they are important in many presentations.)

Employers insist on carefully prepared presentations that completely document all work involved in solving the problems. Thorough documentation may be important in the event of legal considerations, for which the details of the work might be introduced into the court proceedings as evidence. Lack of such documentation may result in the loss of a case that might otherwise have been won. Moreover, internal company use of the work is easier and more efficient if all aspects of the work have been carefully documented and substantiated by data and theory.

Each industrial company, consulting firm, governmental agency, and university has established standards for presenting technical information. These standards vary slightly, but all fall into a basic pattern, which we will discuss. Each organization expects its employees to follow its standards. Details can be easily modified in a particular situation once you are familiar with the general pattern that exists in all of these standards.

It is not possible to specify a single problem layout or format that will accommodate all types of engineering solutions. Such a wide variety of solutions exists that the technique used must be adapted to fit the information to be communicated. In all cases, however, one must lay out a given problem in such a fashion that it can be easily grasped by the reader. No matter what technique is used, it must be logical and understandable.

We have listed guidelines for problem presentation. Acceptable layouts for problems in engineering also are illustrated. The guidelines are not intended as a precise format that must be followed, but, rather, as a suggestion that should be considered and incorporated whenever applicable.

Two methods of problem presentation are typical in the academic and industrial environments. Presentation formats can be either freehand or computer generated. As hardware technology and software developments continue to provide better tools, the use of the computer as a method of problem presentation will continue to increase.

If a formal report, proposal, or presentation is to be the choice of communication, a computer-generated presentation is the correct approach. The example solutions that are illustrated in Figs. 3.1 through 3.4 include both freehand as well as computer output. Check with your instructor to determine which method is appropriate for your assignments. Figure 3.1 illustrates the placement of information.

The following nine general guidelines should be helpful as you develop the freehand skills needed to provide clear and complete problem documentation. The first two examples, Figs. 3.1 and 3.2, are freehand illustrations. The third example, Fig. 3.3, is computer generated with a word processor and Fig. 3.4 uses a spreadsheet to do the computations and graphing.

These guidelines are most applicable to freehand solutions, but many of the ideas and principles apply to computer generation as well.

1. One common type of paper frequently used is called engineering problems paper. It is ruled horizontally, and vertically on the *reverse* side, with only heading and margin rulings on the front. The rulings on the reverse side, which are faintly visible through the paper, help one to maintain horizontal lines of lettering and to provide guides for sketching and simple graph construction. Moreover, the lines on the back of the paper will not be lost as a result of erasures.

2. The completed top heading of the problems paper should include such information as name, date, course number, and sheet number. The upper right-hand block should normally contain a notation such as a/b, where a is the page number of the sheet and b is the total number of sheets in the set.

3. Work should ordinarily be done in pencil using an appropriate lead hardness (HB, F, or H) so that the line work is crisp and unsmudged. Erasures should always be complete, with all eraser particles removed. Letters and numbers must be dark enough to ensure legibility when photocopies are needed.

4. Either vertical or slant letters may be selected as long as they are not mixed. Care should be taken to produce good, legible lettering but without such care that little work is accomplished.

5. Spelling should be checked for correctness. There is no reasonable excuse for incorrect spelling in a properly done problem solution.

6. Work must be easy to follow and uncrowded. This practice contributes greatly to readability and ease of interpretation.

7. If several problems are included in a set, they must be distinctly separated, usually by a horizontal line drawn completely across the page between problems. Never begin a second problem on the same page if it cannot be completed there. Beginning each problem on a fresh sheet is usually better except in cases where two or more problems can be completed on one sheet. It is not necessary to use a horizontal separation line if the next problem in a series begins at the top of a new page.

8. Diagrams that are an essential part of a problem presentation should be clear and understandable. Students should strive for neatness, which is a mark of a professional. Often a good sketch is adequate, but using a straightedge can greatly improve the appearance and accuracy of a diagram. A little effort in preparing a sketch to approximate scale can pay great dividends when it is necessary to judge the reasonableness of an answer, particularly if the answer is a physical dimension that can be seen on the sketch.

9. The proper use of symbols is always important, particularly when the international system (SI) of units is used. It involves a strict set of rules that

Figure 3.1

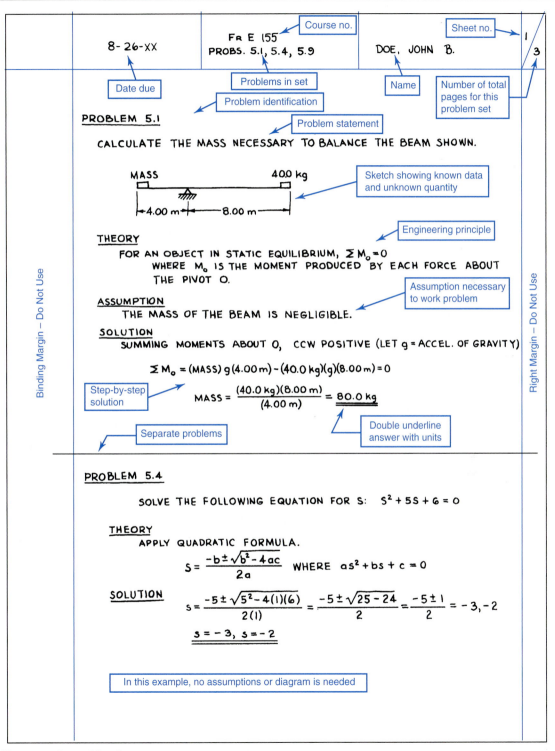

Elements of a problem layout.

Figure 3.2a

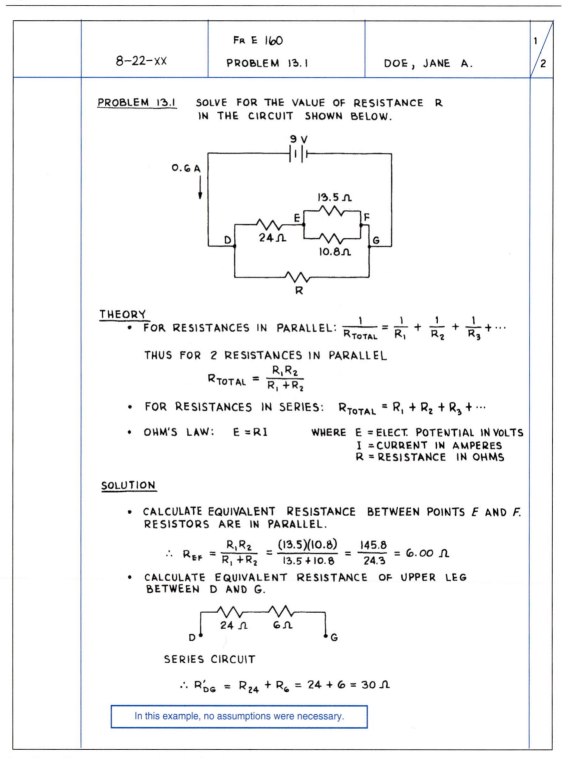

Sample problem presentation done freehand.

Figure 3.2b

- CALCULATE EQUIVALENT RESISTANCE BETWEEN D AND G.

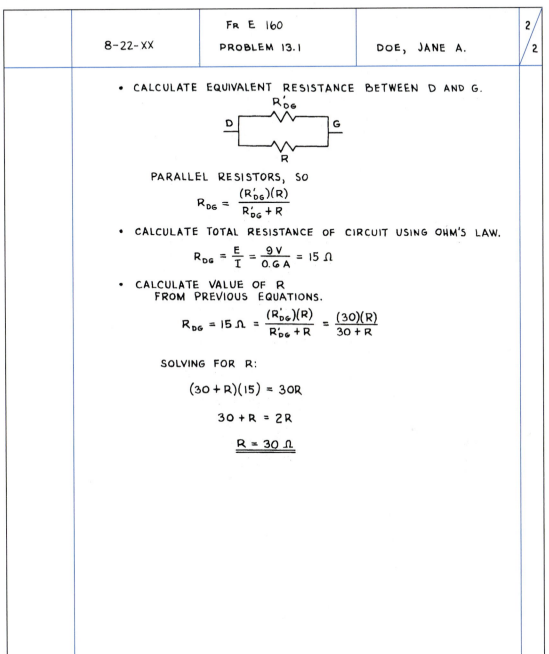

PARALLEL RESISTORS, SO

$$R_{DG} = \frac{(R'_{DG})(R)}{R'_{DG} + R}$$

- CALCULATE TOTAL RESISTANCE OF CIRCUIT USING OHM'S LAW.

$$R_{DG} = \frac{E}{I} = \frac{9\,V}{0.6\,A} = 15\ \Omega$$

- CALCULATE VALUE OF R
  FROM PREVIOUS EQUATIONS.

$$R_{DG} = 15\,\Omega = \frac{(R'_{DG})(R)}{R'_{DG} + R} = \frac{(30)(R)}{30 + R}$$

SOLVING FOR R:

$$(30 + R)(15) = 30R$$

$$30 + R = 2R$$

$$\underline{R = 30\ \Omega}$$

Figure 3.3a

Date   Engineering   Name:_____

Problem

A tank is to be constructed that will hold 5.00×10^5 L when filled. The shape is to be cylindrical, with a hemispherical top. Costs to construct the cylindrical portion will be $300/m^2$, while costs for the hemispherical portion are slightly higher at $400 /m^2$.

Find

Calculate the tank dimensions that will result in the lowest dollar cost.

Theory

Volume of cylinder is... $\qquad V_c = \pi R^2 H$

Volume of hemisphere is... $\qquad V_H = \dfrac{2\pi R^3}{3}$

Surface area of cylinder is... $\qquad SA_c = 2\pi R H$

Surface area of hemisphere is... $\qquad SA_H = 2\pi R^2$

Assumptions

Tank contains no dead air space.
Construction costs are independent of size.
Concrete slab with hermetic seal is provided for the base.
Cost of the base does not change appreciably with tank dimensions.

Solution

1. Express total volume in meters as a function of height and radius.

$$V_{Tank} = f(H, R)$$
$$= V_C + V_H$$
$$500 = \pi R^2 H + \frac{2\pi R^3}{3}$$

Note: $1 m^3 = 1000$ L

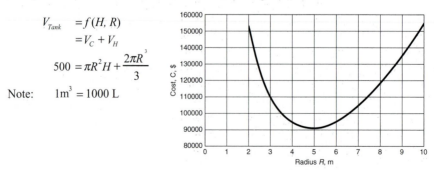

Sample problem presentation done with a word processor.

Figure 3.3b

2. Express cost in dollars as a function of height and radius

$$C = C(H, R)$$

$$= 300 (SA_C) + 400 (SA_H)$$

$$= 300 (2\pi RH) + 400 (2\pi R^2)$$

Note: Cost figures are exact numbers

3. From part 1 solve for $H = H(R)$

$$H = \frac{500}{\pi R^2} - \frac{2R}{3}$$

4. Solve cost equation, substituting $H = H(R)$

$$C = 300 \left[2\pi R \left(\frac{500}{HR^2} - \frac{2R}{3} \right) \right] + 400 \left(2\pi R^2 \right)$$

$$C = \frac{300000}{R} + 400 \pi R^2$$

5. Develop a table of cost versus radius and plot graph.

6. From graph select minimum cost.

$$R = \underline{5.00 \text{ m}}$$
$$C = \$91\ 000$$

7. Calculate H from part 3 above

$$H = \underline{3.033 \text{ m}}$$

8. Verification/check of results from the calculus:

$$\frac{dC}{dR} = \frac{d}{dR} \left[\frac{300\ 000}{R} + 400\pi R^2 \right]$$

$$= \frac{-300\ 000}{R^2} + 800\pi R = 0$$

$$R^3 = \frac{300\ 000}{800\pi}$$

$$R = \underline{4.92\text{m}}$$

Cost versus Radius

Radius, R, m	Cost, C, \$
1.0	301 257
2.0	155 027
3.0	111 310
4.0	95 106
5.0	91 416
6.0	95 239
7.0	104 432
8.0	117 925
9.0	135 121
10.0	155 664

Figure 3.4

Problem 3-5

Analyze the buckling load for steel columns ranging from 50 to 100 ft long in increments of 5 ft.
The cross-sectional area is 7.33 in², the least radius of gyration is 3.19 in and modulus of elasticity is 30 × 10⁶ lb/in².
Plot the buckling load as a function of column length for hinged ends and fixed ends.

Theory

Euler's equation gives the buckling load for a slender column.

$$F_B = \frac{n\pi^2\, EA}{(L/r)^2}$$

where

F_B = buckling load, lb
E = modulus of elasticity, lb/in²   3.00E+07
A = cross-sectional area, in²   7.33
L = length of column, in
r = least radius of gyration, in   3.19
The factor n depends on the end conditions: If both ends are hinged, $n = 1$;
if both ends are fixed, $n = 4$; if one end is fixed and the other is hinged, $n = 2$

Assumption: The columns being analyzed meet the slenderness criterion for Euler's equation

Solution

Length, ft	Buckling load (fixed), lb	Buckling load (hinged), lb
50	245394	61348
55	202805	50701
60	170412	42603
65	145204	36301
70	125201	31300
75	109064	27266
80	95857	23964
85	84911	21228
90	75739	18935
95	67976	16994
100	61348	15337

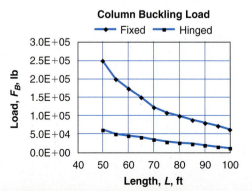

Sample problem presentation done with a spreadsheet.

must be followed so that absolutely no confusion of meaning can result. There also are symbols in common and accepted use for engineering quantities that can be found in most engineering handbooks. These symbols should be used whenever possible. It is important that symbols be consistent throughout a solution and that all of them are defined for the benefit of the reader and for your own reference.

The physical layout of a problem solution logically follows steps that are similar to those of the engineering method. You should attempt to present the process by which the problem was solved in addition to the solution so that any reader can readily understand all the aspects of the solution. Figure 3.1 illustrates the placement of the information.

Figures 3.2, 3.3, and 3.4 are examples of typical engineering problem solutions. You may find these examples to be helpful guides as you prepare your problem presentations.

Key Terms and Concepts

The following are some of the terms and concepts you should recognize and understand.

Process	Problem presentation
Analysis	Solution documentation
Engineering method	

Problems

The solution to lengths and angles of oblique triangles can be arrived at by the application of fundamental trigonometry. All angles are to be considered precise numbers. Solve the following problems using Fig. 3.5 as a general guide.

3.1. Given one side and two angles of an oblique triangle.

$$C = 40.0° \qquad a = 7.50 \text{ m (meters)} \qquad A = 110.0°$$

Using the law of sines and the sum of angles, determine angle B and sides b and c of the triangle.

3.2. Given two sides and the included angle of an oblique triangle.

$$A = 124°$$

$$b = 5.23 \text{ cm (centimeters)}$$

$$c = 8.79 \text{ cm}$$

Figure 3.5

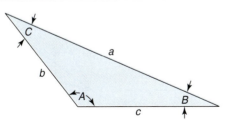

Using the law of sines, the law of cosines, and the sum of angles, find angles B and C and the length of side a.

3.3 Given the sides of an oblique triangle

$$a = 11.1 \text{ in (inches)}$$

$$b = 9.35 \text{ in}$$

$$c = 6.79 \text{ in}$$

determine the angles A, B, and C to the nearest tenth of a degree using the law of cosines and the sum of angles.

3.4 Given two sides and the included angle of an obtuse triangle

$$C = 25.5°$$

$$b = 6.66 \times 10^3 \text{ km (kilometers)}$$

$$a = 16\ 300 \text{ km}$$

(a) Using the sum of angles, the law of sines, and the law of tangents, find the missing angles and sides.
The law of tangents is

$$\frac{a + b}{a - b} = \frac{\tan\left(\dfrac{A + B}{2}\right)}{\tan\left(\dfrac{A - B}{2}\right)}$$

(b) Find the radius of an inscribed circle (r) and the circumscribed circle (R).
Vectors have both magnitude and direction. Figure 3.6 applies to Probs. 3.5 through 3.7 with positive angles shown counterclockwise from the positive x axis.

3.5 For Fig. 3.6a, find
(a) the vertical (A_y) and horizontal (A_x) components of A if vector **A** has a magnitude of 523 N (newtons) and angle $\phi = 24°$.
(b) the magnitude and direction of **A** if $A_x = 72.3$ m and $A_y = 47.7$ m.

3.6 For Fig. 3.6b, find
(a) the magnitude of current V_R when:

$$\text{current } V_1 = 13 \text{ knots} \qquad \alpha = 17° \qquad \delta = 13°$$

$$\text{current } V_2 = 10 \text{ knots} \qquad \beta = 19°$$

Figure 3.6

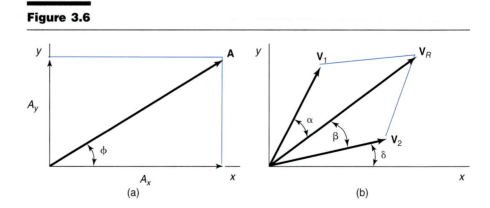

(a)   (b)

Figure 3.7

139
Problems

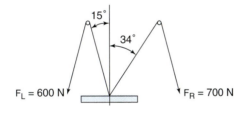

(b) the magnitude of V_2 and angle α when:

$$V_R = 476 \text{ ft/s (feet/second)} \qquad \delta = 5°$$
$$V_1 = 217 \text{ ft/s} \qquad \beta = 27°$$

3.7 Find the resultant magnitude (V_T) and the direction of the sum of the three vectors in Fig. 3.6b when

$$V_R = -1.32 \text{ mm} \qquad \alpha = 57°$$
$$V_1 = 2.31 \text{ mm} \qquad \beta = 89°$$
$$V_2 = 0.78 \text{ mm} \qquad \delta = 21°$$

3.8 Two workers are trying to lift a slab of concrete. They are using ropes passing through pulleys suspended above the slab as shown in Fig. 3.7. If the person on the left can pull with a force of 600 N and the worker on the right can pull with a force of 700 N, what mass of concrete can they lift together?

3.9 Jack and Jill were flying cross country in their single-engine plane when their engine quit. When the engine died, the altimeter read 800 ft. They had just passed over a radio tower used as a marker on their navigation map. According to the map, 2 mi (miles) straight ahead was a small private airport. If the glide ratio of the plane is 18 : 1, will they be able to land safely at the airport?

3.10 How fast did the lunar landing module travel when it was sitting on the surface of the moon if you consider only its orbital motion around the earth in kilometer per hour? Assume a radius of 3.84×10^6 m from the earth and a time of 27.3 days to complete the orbit. (The diameter of the earth may be disregarded.)

3.11 A group of engineering students decided to climb a mountain over spring break. After walking for 6 hours they came to a flat clearing with a sign that said elevation 987 ft. Since they were very bright engineering students, they decided to determine how much farther they had to go. They measured the height of a small pine tree and found it to be 8 ft tall. Then one of them laid on the ground and lined up the peak of the mountain with the top of the tree. They measured the distance between the student and the tree and found it to be 13 ft. They knew the height of the peak was 1347 ft. How much farther did they have to go to reach the top of the mountain?

3.12 Jill wants to practice swimming for a triathlon. She wants to know how far it is from the dock of her pond to the marker on the opposite bank. According to her map, the pond sits between two roads 1 mi apart north and south and two roads 2 mi apart east and west. She has measured the angles as shown in Fig. 3.8. Find the distance between the two points and determine how many times she must swim over and back to swim at least 3 km (kilometers).

3.13 A certain polymer needs to be heated by a heat lamp for at least 45 s (seconds) to cure properly at 50° C at 1 atm pressure. The optimum height of the heat lamp

Figure 3.8

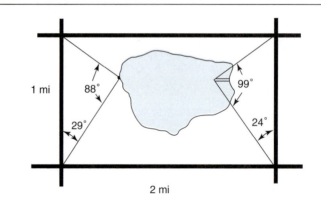

1 mi   88°   99°

29°   24°

2 mi

above the parts is 75 cm. The heat lamp has an effective range within a circle 25°
from straight down.

(a) If the conveyor belt is moving at 5 m/min (meters per minute), how many
heat lamps must be placed in succession above the parts to cure properly?

(b) If three heat lamps are used, what should the speed of the belt be?

3.14 An engineering student has been hired by a machining house to help program
CNC lathes. The part that needs to be machined ranges from a starting diameter
of 2.5 in to a finished diameter of 0.425 in on one end. The part to be machined
is stainless steel with a recommended cutting speed of 400 fpm (feet per minute)
for the carbide roughing cutter and 350 fpm for the finish cutter.

(a) How many rpm (revolutions per minute) should the engineering student
start the program for the 2.5 in diameter, and what should the spindle speed
be for the final finish cut in the small diameter?

(b) If the material is changed to brass with a cutting speed of 300 fpm and the
whole part run at a single spindle speed, what speed should it be?

3.15 John, an engineering student, is on vacation and standing on Mt. San Jacinto,
looking down on Palm Springs, whose elevation is 487 ft above sea level. John is
standing 8516 ft above sea level. It is a clear day so John can easily see 30 mi in a
semicircle from his viewpoint. How many acres can John see from where he is
standing? Be sure to include your assumptions with your answer.

3.16 The drive pulley for a conveyor belt is 25.0 cm in diameter. The pulley rotates
2.50 times per second. Express this in (a) radians per minute and (b) revolutions
per minute. (c) If the idler wheel at the other end of the conveyor belt is 10.0 cm
in diameter, how fast is it turning? (d) What is the speed of the belt in meters per
minute?

3.17 In the 2000 summer Olympics swimmers wore swimming suits made from a
revolutionary new material that had less drag in water than bare skin. If a
swimmer with 1.600 m^2 surface area wore a suit with 550.0 cm^2 and a drag factor
of 2.500 percent above that of bare skin swam the 100 m freestyle in 48.92 s, how
fast could he swim the same event in a suit with 1.000×10^4 cm^2 and a 1.000 percent reduction in drag from bare skin. Assume that all other factors are constant.

3.18 The foundation for a large office building requires footings of different thicknesses as shown in Fig. 3.9. These footings will be poured the full length of the
building plus an additional 5 ft 0 in on each end. The building will be 110'0" long.

(a) If the outside walls are to be 1'8" thick and centered on the 12' pads, what is
the outside width of the building to the nearest foot?

Figure 3.9

141

Problems

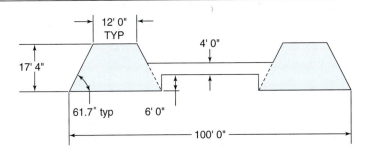

(b) How many yards (cubic yards) of concrete should the contractor order for the footings? [Give your answer to the nearest 10 yd (yards).]

3.19 For Prob. 3.18, if a 1:3:4 mix is used with 5.5 gal (gallons) of water for each bag of Portland cement, this means that 1 ft^3 (cubic foot) of Portland cement (94 lb) is combined with 3 ft^3 of fine sand [105 lb/ft^3 (pounds per cubic foot)] and 4 ft^3 of rock (100 lb/ft^3) and 5.5 gal of water. (1 gal water = 8.33 lb.) Assume no loss due to handling and 27 ft^3 of the first three materials to yield 1 yd^3 (cubic yard) of concrete. (Do not include the weight of the water.)

(a) What would the total weight of the footings be in tons?

(b) What is the amount of each material and the water used in the footings to the nearest ton?

3.20 A construction company is looking into purchasing the plot of land shown in Fig. 3.10 and dividing it into 2-acre lots for a housing development. The parcel is formed by the intersection of three roads and a lake on the right. Payne Road intersects US 19 at an angle of 58° and US 19 intersects Green Belt Drive at 98°. The distance between points A and B is 875 yd. The distance between points B and C is 1000 yd and the distance between points C and D is 955 yd.

(a) Find the distance between points A and D in feet.

(b) Find the total area in square feet of the land bordered by the four points disregarding the irregular coastline.

(c) Determine how many 2-acre plots can be sold in the development.

Figure 3.10

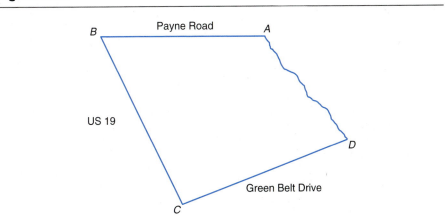

Figure 3.11

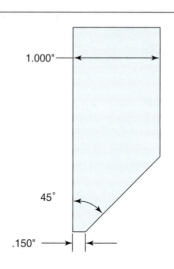

(d) If the lots bordering on the lake have a minimum of a 200 foot frontage and sell for $65,000 each and the other lots all sell for $43,000, what is the total value of the lots in the development?

3.21 A tooling designer is designing a jig that will insert pins into the flip-up handles of a coffeemaker. The bin to hold the pins should hold enough pins for 3 hours' work before reloading. The engineering estimate for the rate of inserting the pins is 300 per hour. The pin diameter is 0.125 in. Because of space constraints, the bin must be designed as shown in Fig. 3.11. The bin only can be the width of one pin. What is the minimum height that the bin should be if there is 0.250 in at the top when it has been filled with the required number of pins?

3.22 Two friends are planning to go on RAGBRAI (Register's Annual Great Bike Ride Across Iowa) next year. One is planning to ride a mountain bicycle with 26 in tires and the other has a touring bicycle with 27 in tires. A typical RAGBRAI is about 480 mi long.

(a) How many more revolutions will the mountain bike tires make in that distance than the touring bike?

(b) Typical gearing for most bicycles ranges from 30 to 50 teeth on the chain-wheel (front gears) and 12 to 30 teeth on the rear cog. Find how many more revolutions of the pedals the mountain biker will make during the trip, using an average pedaling time of 85 percent of the trip and a 42 tooth chainwheel and a 21 tooth rear cog for your calculations.

(c) If both cyclists have 170 mm cranks on their bikes, what will be the mechanical advantage (considering only the movement of the feet with a constant force for walking and riding), in percent, that each rider will have achieved over walking the same distance? (This can be found by dividing the distance walked by the distance the feet move in riding.)

3.23 An engineer has been given the assignment of finding how much money can be saved over a year's time by redesigning the press plates from the pattern shown in Fig. 3.12a to the pattern shown in Fig. 3.12b for stamping out 2.400 inch diameter disks. The stamping material is 14 gage sheet metal and can be purchased in 100 foot rolls in varying widths in one-half inch increments. One square foot of metal weighs 3.20 pounds. The metal is sold for $0.20 a pound. Do

Figure 3.12

143

Problems

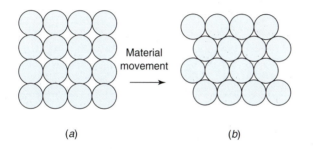

Material
movement

(a)   (b)

not consider the ends of the rolls. The company expects to produce 38 000 parts this year. How much can be saved?

3.24 Sally is making a sine bar, which is used to machine angles on parts. See Fig. 3.13. She has a 1.250 in thick bar which needs 90° grooves machined into it for precision ground 1.0000 in. diameter cylinders A sine bar is used by placing different thicknesses under one of the cylinders so that the proper angle is attained. Sally wants the distance between the centers of the cylinders to be 5.000 in.

(a) How deep should she mill the 90° grooves so that the top of the sine block is 2 in tall?

(b) Once her sine block is finished, she wants to mill a 22.5° angle on a brass block. What thickness of gage blocks will produce this angle for this sine plate?

3.25 Standing at the edge of the roof of a tall building you throw a ball upward with a velocity of 15 m/s (meters per second). The ball goes straight up and begins its downward descent just missing the edge of the building. The building is 40 m tall.

(a) What is the velocity of the ball at its uppermost position?

(b) How high above the building will the ball go before beginning its descent?

(c) What will be the velocity of the ball as it passes the roof of the building?

(d) What will be the speed of the ball just before it hits the ground?

(e) How long will it take for the ball to hit the ground after leaving your hand?

3.26 A stuntperson is going to attempt a jump across a canyon 74 m wide. The ramp on the far side of the canyon is 25 m lower than the ramp from which she will leave. The takeoff ramp is built with a 15° angle from horizontal.

(a) If the stuntperson leaves the ramp with a velocity of 28 m/s, will she make the jump?

(b) How many seconds will she be in the air?

Figure 3.13

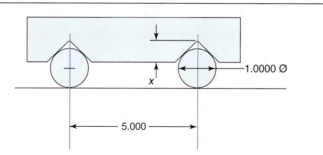

1.0000 Ø

x

5.000

Figure 3.14

3.27 An engineering student has been given the assignment of designing a hydraulic holding system for a hay baling system. The system has four cylinders with 120 mm diameter pistons with a stroke of 0.320 m. The lines connecting the system are 1 cm id (inside diameter). There are 15.5 m of lines in the system. For proper design the reserve tank should hold a minimum of 50 percent more than the amount of hydraulic fluid in the system. If the diameter of the reserve tank is 30.48 cm, what is the shortest height it should be?

3.28 The plant engineer for a large foundry has been asked to calculate the thermal efficiency of the generating plant used by the company to produce electricity for the aluminum melting furnaces. The plant generates 545.6 GJ of electrical energy daily. The plant burns 50 t (tons) of coal a day. The heat of combustion of coal is about 6.2×10^6 J/kg (Joules/kilogram). What was the answer? (Efficiency = W/J_{heat})

3.29 Using these three formulas,

$$V = IR \qquad R = (\rho L)/A \qquad A = \pi(0.5d)^2$$

Find the difference in current (I) that a copper wire ($\rho = 1.72 \times 10^{-8}\ \Omega \cdot$ m) can carry over an aluminum wire ($\rho = 2.75 \times 10^{-8}\ \Omega \cdot$ m) with equal diameters (d) of 0.5 cm and a length (L) of 10 000 m carrying 110 V (volts).

3.30 The light striking a pane of glass is refracted as shown in Fig. 3.14. The law of refraction states that $n_a \sin \theta_a = n_b \sin \theta_b$, where n_a and n_b are the refraction indexes of the materials through which the light is passing and the angles are from a line that is normal to the surface. The refractive index of air is 1.00. What is the refractive index of the glass?

CHAPTER 4

Representation of Technical Information

4.1 Introduction

This chapter begins with an example of an actual freshman engineering student team project. This team consisted of aerospace, electrical, and mechanical engineering students who were assigned to find how temperature and pressure varied with altitude. The team was not given specific instructions as to how this might be accomplished, but once the information had been collected, they were expected to record, plot, and analyze the data.

After a bit of research the team decided to request a university plane that was equipped with the latest Rockwell Collins avionics gear, and since it was a class assignment, the team members asked if the university's Air Flight Service would consider helping them to conduct this experiment "free of charge." Since the pilots do periodic maintenance flights, they agreed to allow the students to ride along.

Using the plane's sophisticated data acquisition equipment, the students were able to collect and record the data needed for the assignment (see Table 4.1) along a flight path as the plane ascended to 15 000 feet. They decided that one student would record, one would make temperature readings, and the third would take pressure readings.

Table 4.1

Height, H, ft	Temperature, T, °F	Pressure, P, lbf/in²
0	59	14.7
1 000	55	14.2
2 000	52	13.7
3 000	48	13.2
4 000	44	12.7
5 000	41	12.2
6 000	37	11.8
7 000	34	11.3
8 000	30	10.9
9 000	27	10.5
10 000	23	10.0
11 000	19	9.7
12 000	16	9.3

Figure 4.1

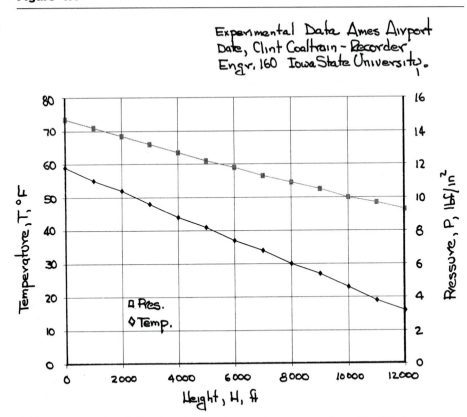

Experimental Data Ames Airport
Data, Clint Coaltrain - Recorder,
Engr. 160 Iowa State University.

Table constructed recording how temperature and pressure varies with altitude.

When the students returned to the hangar, they made a freehand plot (with straightedge) of the collected information, see Fig. 4.1.

Since the plot demonstrated linear results, the students returned to campus to prepare the required report, including an analysis of the data collected.

They decided that the written report should be supported with a computer-generated table, a computer-generated plot of the data, and a computer-generated least-squared curve fit to determine the mathematical relationship between the variables.

Table 4.2 represents a computer-generated table once the students returned to campus. This could be done with word processing, spreadsheet, or other commerially available software packages.

Figures 4.2 and 4.3 represent spreadsheet applications that are powerful and extremely convenient once the fundamentals of good graph construction are understood. Figure 4.2 is an example of an EXCEL spreadsheet using a scatter plot with each data point connected with a straight line. Figure 4.3 is an example of a scatter plot with only the data points plotted. A trendline is then applied using the method of least squares with the equation of the line included. This concept will be covered later in this chapter. Both of these plots required considerable manipulation of the software to arrive at the finished

Table 4.2

Height, *H*, ft	Temperature, *T*, °F	Pressure, *P*, lbf/in^2
0	59	14.7
1 000	55	14.2
2 000	52	13.7
3 000	48	13.2
4 000	44	12.7
5 000	41	12.2
6 000	37	11.8
7 000	34	11.3
8 000	30	10.9
9 000	27	10.5
10 000	23	10.0
11 000	19	9.7
12 000	16	9.3

Figure 4.2

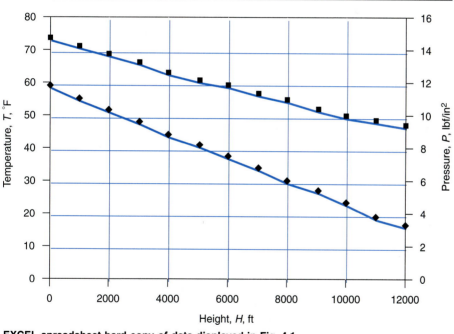

EXCEL spreadsheet hard copy of data displayed in Fig. 4.1.

products. Therefore it is very important to understand all the material presented in this chapter so that when you elect to use commercially available packages, your end results will be displayed and presented correctly.

This chapter contains examples and guidelines as well as helpful information that will be needed when collecting, recording, plotting, and interpreting technical data. Two areas will be considered in considerable detail: (1) graphical presentation of scientific data and (2) graphical analysis of plotted data.

Figure 4.3

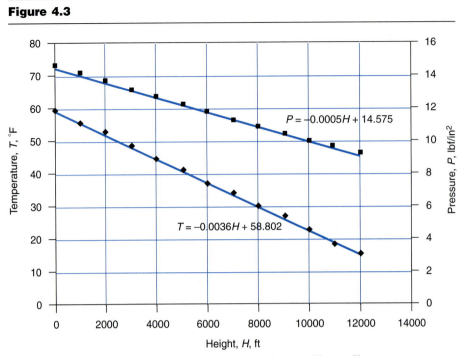

$$P = -0.0005H + 14.575$$

$$T = -0.0036H + 58.802$$

EXCEL spreadsheet hard copy of data displayed in Fig. 4.1 with trendline.

Graphical presentation of technical data is necessary when calculated or experimental results are recorded in tabular form. Rapid and accurate determination of relationships between numerical values when the information is reported in columns and rows is difficult. A procedure for graphing the results is needed. This approach provides a visual impression that is a more intuitive method with which to compare variables, rates of change, or relative magnitudes.

However, complete graphical analysis involves correct and accurate interpretation of data after they have been plotted. At times impressions are not sufficient, so determination of a mathematical model is required.

Computers, with their expanding power, versatility, and speed, are changing the way we collect, record, display, and analyze data. For example, consider the rapidly rising popularity of the microprocessor. It has changed the way we do many tasks. Microprocessors or small CPU (central processing unit) cards can be programmed by a remote host computer, normally a PC. These CPUs are interfaced with some device, perhaps a temperature sensor or a robot. Information can be collected and stored and then, based on the feedback of the information to the CPU, additional programmed instructions can direct the device to perform corrective actions.

In addition to microprocessors and various other types of data collection instrumentation, there exists a wide variety of commercial software that enables the engineer to reduce the time required for recording, plotting, and analyzing data while increasing the accuracy of the results. Numerous software programs are available for both technical presentation as well as analysis. These programs provide a wide range of powerful tools.

4.1.1 Software for Recording and Plotting Data

Data are recorded in the field as shown in Table 4.1. Many times a quick, free-hand plot of the data is produced to provide a visual impression of the results while still in the field (see Fig. 4.1).

Upon returning to the laboratory, however, you will find that spreadsheet software, such as EXCEL and LOTUS 1-2-3, provide enormous recording and plotting capability. The data are entered into the computer and by manipulation of software options both the data and a graph of the data can be configured, stored, and printed (Table 4.2 and Figs. 4.2 and 4.3).

Programs, such as Mathematica, Matlab, Mathcad, TK Solver, and many others, provide a range of powerful tools designed to help analyze numerical and symbolic operations as well as present a visual image of the results.

Software also is widely available to provide methods of curve fitting once the data have been collected and recorded. This subject will be treated in a separate chapter.

Even though it is important for the engineer to interpret, analyze, and communicate different types of data, it is not practical to include in this chapter all the forms of graphs and charts that may be encountered. For that reason popular appeal or advertising charts such as bar charts, pie diagrams, and distribution charts, although useful to the engineer, will not be discussed here.

Even though commercial software is extremely helpful during the presentation and analysis process, the results are only as good as the original software design and its use by the operator. Some software provides a wide range of tools but only allows limited data applications and minimal flexibility to modify default outputs. Other software provides a high degree of in-depth analysis for a particular subject area with considerable latitude to adjust and modify parameters.

Inevitably the computer together with a growing array of software will continue to provide an invaluable analysis tool. However, it is absolutely essential that users be extremely knowledgeable of the software and demonstrate considerable care when manipulating the data. They need to understand the software limitations and accuracies, but, above all, the operator must know what plotted results are necessary and what expectations for appearance and readability are mandated.

For this reason the sections to follow are a combination of manual collection, recording, plotting, and analysis and computer-assisted collection, recording, plotting, and analysis. As we begin the learning process, it is important to know how to manipulate data mechanically so that computer results can be appropriately modified.

4.2 Collecting and Recording Data

Modern science was founded on scientific measurement. Meticulously designed experiments, carefully analyzed, have produced volumes of scientific data that have been collected, recorded, and documented. For such data to be meaningful, however, certain laboratory procedures must be followed. Formal data sheets, such as those shown in Fig. 4.4, or laboratory notebooks should be

Figure 4.4

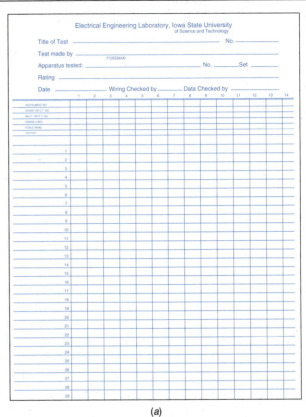

(a)

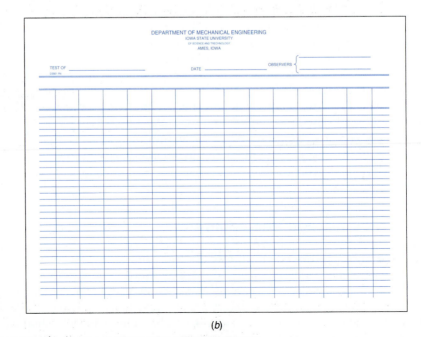

(b)

Sample data sheets used by engineering departments.

used to record all observations. Information about equipment, such as the instruments and experimental apparatus used, should be recorded. Sketches illustrating the physical arrangement of equipment can be very helpful. Under no circumstances should observations be recorded elsewhere or data points erased. The data sheet is the "notebook of original entry." If there is reason for doubting the value of any entry, it may be canceled (i.e., not considered) by drawing a line through it. The cancellation should be done in such a manner that the original entry is not obscured, in case you want to refer to it later.

Sometimes a measurement requires minimal precision, so that time can be saved by making rough estimates. As a general rule, however, it is advantageous to make all measurements as precisely as time and the economics of the situation will allow. Unfortunately, as different observations are made throughout any experiment, some degree of inconsistency will develop. Errors enter into all experimental work regardless of the amount of care exercised.

It can be seen from what we have just discussed that the analysis of experimental data involves not only measurements and collection of data but also careful documentation and interpretation of results.

4.2.1 Manual Entry

Experimental data once collected are normally organized into some tabular form, which is the next step in the process of analysis. Data, such as that shown in Table 4.1, should be carefully labeled and neatly lettered so that results are not misinterpreted. This particular collection of data represents atmospheric pressure and temperature measurements recorded at various altitudes by students during a flight in a light aircraft.

Although the manual tabulation of data is frequently a necessary step, you will sometimes find it difficult to visualize a relationship between variables when simply viewing a column of numbers. A most important step in the sequence from collection to analysis therefore, is the construction of appropriate graphs or charts.

4.2.2 Computer Assisted

In recent years, a variety of equipment has been developed which will automatically sample experimental data for analysis. We expect to see expansion of these techniques along with continuous visual displays that will allow us interactively to control the experiments. As an example the flight data collected onboard the aircraft could be entered directly into a laptop computer and printed as in Table 4.2.

4.3 General Graphing Procedures

Many examples will be used throughout this chapter to illustrate methods of graphical presentation because their effectiveness depends to a large extent on the details of construction.

The proper manual construction of a graph from tabulated data can be generalized into a series of steps. Each of these steps will be discussed and illustrated in considerable detail in the following sections. Once you understand the entire manual process of graph construction, the step to computer-generated graphs will be simple; see Sec. 4.3.8.

1. Select the type of graph paper (rectilinear, semilog, log-log) and grid spacing for best representation of the given data.
2. Choose the proper location of the horizontal and vertical axes.
3. Determine the scale units (range) for each axis to display the data appropriately.
4. Graduate and calibrate the axes using the 1, 2, 5 rule.
5. Identify each axis completely.
6. Plot points using permissible symbols.
7. Check any point that deviates from the slope or curvature of the line.
8. Draw the curve or curves.
9. Identify each curve, add title and necessary notes.
10. Darken lines for good reproduction.

4.3.1 Graph Paper

Printed coordinate graph paper is commercially available in various sizes with a variety of grid spacing. Rectilinear ruling can be purchased in a range of lines per inch or lines per centimeter, with an overall paper size of 8.5 × 11 inches considered most typical.

Closely spaced coordinate ruling is generally avoided for results that are to be printed or photoreduced. However, for accurate engineering analyses requiring some amount of interpolation, data are normally plotted on closely spaced, printed coordinate paper. Graph paper is available in a variety of colors, weights, and grades. Translucent paper can be used when the reproduction system requires a material that is not opaque.

If the data require the use of log-log or semilog paper, such paper also can be purchased in different formats, styles, weights, and grades. Both log-log and semilog grids are available from 1 to 5 cycles per axis. (A later section will discuss different applications of log-log and semilog paper.). Examples of commercially available log-log and semilog paper are given in Figs. 4.5a and b.

4.3.2 Axes Location and Breaks

The axes of a graph consist of two intersecting straight lines. The horizontal axis, normally called the *x-axis*, is the *abscissa*. The vertical axis, denoted by the *y-axis*, is the *ordinate*. Common practice is to place the independent values along the abscissa and the dependent values along the ordinate, as illustrated in Fig. 4.6.

Many times mathematical graphs contain both positive and negative values of the variables. This necessitates the division of the coordinate field into four quadrants, as shown in Fig. 4.7. Positive values increase toward the right and upward from the origin.

On any graph a full range of values is desirable, normally beginning at zero and extending slightly beyond the largest value. To avoid crowding, one should use the entire coordinate area as completely as possible. However, certain circumstances require special consideration to avoid wasted space. For example, if values to be plotted along the axis do not range near zero, a "break" in the grid or the axis may be used, as shown in Figs. 4.8a and b.

Figure 4.5

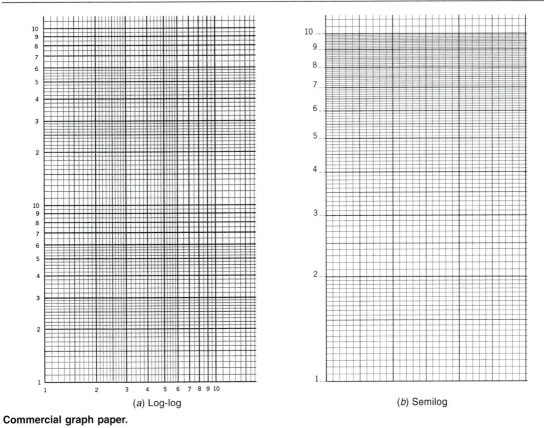

(a) Log-log

(b) Semilog

Commercial graph paper.

Figure 4.6

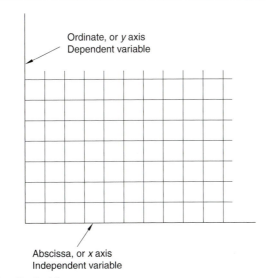

Ordinate, or *y* axis
Dependent variable

Abscissa, or *x* axis
Independent variable

Abscissa (*x*) and ordinate (*y*) axes.

Figure 4.7

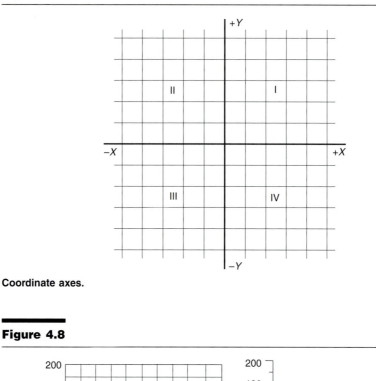

Coordinate axes.

Figure 4.8

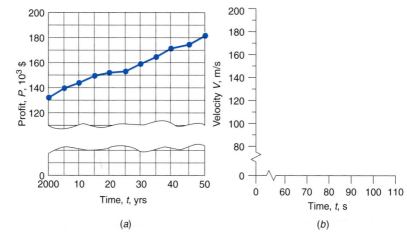

(a)

(b)

Typical axes breaks.

When judgments concerning relative amounts of change in a variable are required, the axis or grid should not be broken or the zero line omitted, with the exception of time in years, such as 1970, 1971, and so on, since that designation normally has little relation to zero.

Since most commercially prepared grids do not include sufficient border space for proper labeling, the axes should preferably be placed 20 to 25 mm (approximately 1 in) inside the edge of the printed grid in order to allow ample room for graduations, calibrations, axes labels, reproduction, and binding.

The edge of the grid may have to be used on log-log paper since it is not always feasible to move the axis. However, with careful planning the vertical and horizontal axes can be repositioned in most cases, depending on the range of the variables.

4.3.3 Scale Graduations, Calibrations, and Designations

The scale is a series of marks, called *graduations,* laid down at predetermined distances along the axis. Numerical values assigned to significant graduations are called *calibrations.*

A scale can be uniform, with equal spacing along the stem, as found on the metric, or engineer's scales. If the scale represents a variable whose exponent is not equal to 1 or a variable that contains trigonometric or logarithmic functions, the scale is called a *nonuniform,* or *functional scale.* Examples of both these scales together with graduations and calibrations are shown in Fig. 4.9. When you plot data, one of the most important considerations is the proper selection of scale graduations. A basic guide to follow is the 1, 2, 5 rule, which can be stated as follows:

> Scale graduations are to be selected so that the smallest division of the axis is a positive or negative integer power of 10 times 1, 2, or 5.

The justification and logic for this rule are clear. Graduation of an axis by this procedure makes possible the interpolation of data between graduations when plotting or reading a graph. Figure 4.10 illustrates both acceptable and nonacceptable examples of scale graduations.

Figure 4.9

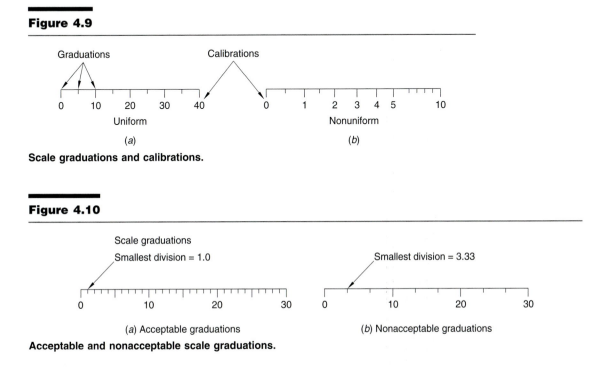

(a) Uniform — Graduations — 0 10 20 30 40

(b) Nonuniform — Calibrations — 0 1 2 3 4 5 10

Scale graduations and calibrations.

Figure 4.10

Scale graduations
Smallest division = 1.0

(a) Acceptable graduations — 0 10 20 30

Smallest division = 3.33

(b) Nonacceptable graduations — 0 10 20 30

Acceptable and nonacceptable scale graduations.

Figure 4.11

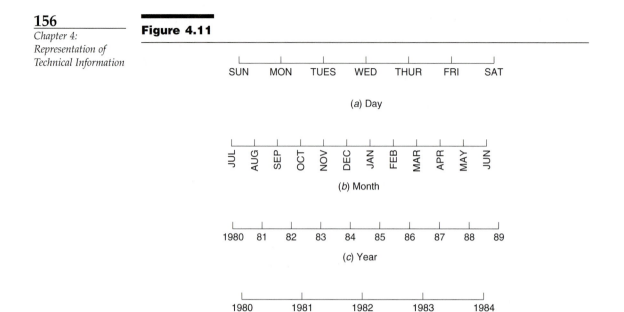

(a) Day

(b) Month

(c) Year

(d) Year

Time as a variable.

Violations of the 1, 2, 5 rule that are acceptable involve certain units of time as a variable. Days, months, and years can be graduated and calibrated as illustrated in Fig. 4.11.

Scale graduations normally follow a definite rule, but the number of calibrations to be included is primarily a matter of good judgment. Each application requires consideration based on the scale length and range as well as the eventual use. Figure 4.12 demonstrates how calibrations can differ on a scale with the same length and range. Both examples obey the 1, 2, 5 rule, but as you can see, too many closely spaced calibrations make the axis difficult to read.

The selection of a scale deserves attention from another point of view. If the rate of change is to be depicted accurately, then the slope of the curve should represent a true picture of the data. By contracting or expanding the axis or axes, you could imply an incorrect impression of the data. Such a procedure is to be avoided. Figure 4.13 demonstrates how the equation $Y = X$ can be misleading if not properly plotted. Occasionally distortion is desirable, but it should always be carefully labeled and explained to avoid misleading conclusions.

If plotted data consist of very large or small numbers, the SI prefix names (milli-, kilo-, mega-, etc.) may be used to simplify calibrations. As a guide, if the numbers to be plotted and calibrated consist of more than three digits, it is customary to use the appropriate prefix; an example is illustrated in Fig. 4.14.

The length scale calibrations in Fig. 4.14 contain only two digits, but the scale can be read by understanding that the distance between the first and second graduation (0 to 1) is a kilometer (km); therefore the calibration at 10 represents 10 km.

Figure 4.12

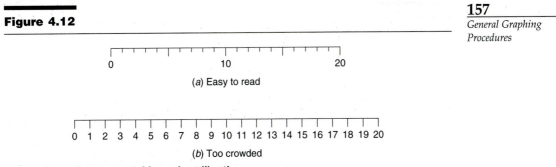

(a) Easy to read

(b) Too crowded

Acceptable and nonacceptable scale calibrations.

Figure 4.13

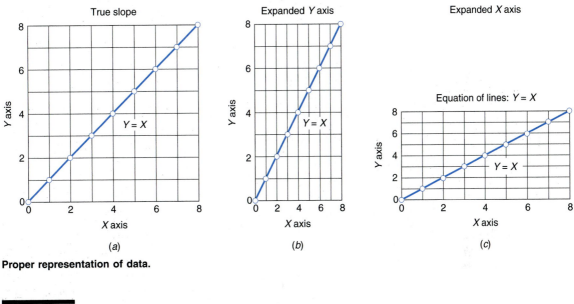

(a)

(b)

(c)

Proper representation of data.

Figure 4.14

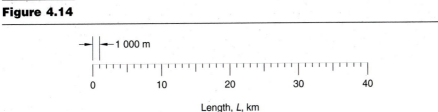

Length, L, km

Reading the scale.

Certain quantities, such as temperature in degrees Celsius and altitude in meters, have traditionally been tabulated without the use of prefix multipliers. Figure 4.15 depicts a procedure by which these quantities can be conveniently calibrated. Note in particular that the distance between 0 and 1 on the scale represents 1000°C. This is another example of how the SI notation is convenient

Figure 4.15

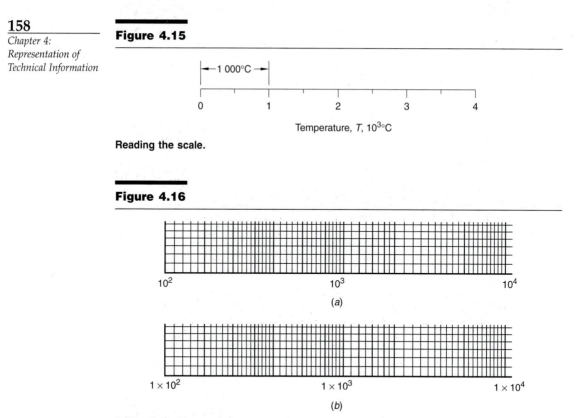

Reading the scale.

Figure 4.16

Calibration of log scales.

since the prefix multipliers (micro-, milli-, kilo-, mega-, etc.) allow the calibrations to stay within the three-digit guideline.

The calibration of logarithmic scales is illustrated in Fig. 4.16. Since log-cycle designations i.e., start and end with powers of 10 (i.e., 10^{-1}, 10^0, 10^1, 10^2, etc.), and since commercially purchased paper is normally available with each cycle printed 1 through 10, Figs. 4.16a and b demonstrate two preferred methods of calibration.

4.3.4 Axis Labeling

Each axis should be clearly identified. At a minimum the axis label should contain the name of the variable, its symbol, and its units. Since time is frequently the independent variable and is plotted on the x axis, it has been selected as an illustration in Fig. 4.17. Scale designations should preferably be placed outside

Figure 4.17

Axis identification.

the axes, where they can be shown clearly. Labels should be lettered parallel to the axis and positioned so that they can be read from the bottom or right side of the page as illustrated in Fig. 4.22.

159

General Graphing Procedures

4.3.5 Point-Plotting Procedure

Data can normally be categorized in one of three general ways: as observed, empirical, or theoretical. Observed and empirical data points are usually located by various symbols, such as a small circle or square around each data point, whereas graphs of theoretical relations (equations) are normally constructed smooth, without use of symbol designation. Figure 4.18 illustrates each type.

4.3.6 Curves and Symbols

On graphs prepared from observed data resulting from laboratory experiments, points are usually designated by various symbols (see Fig. 4.19). If more than

Figure 4.18

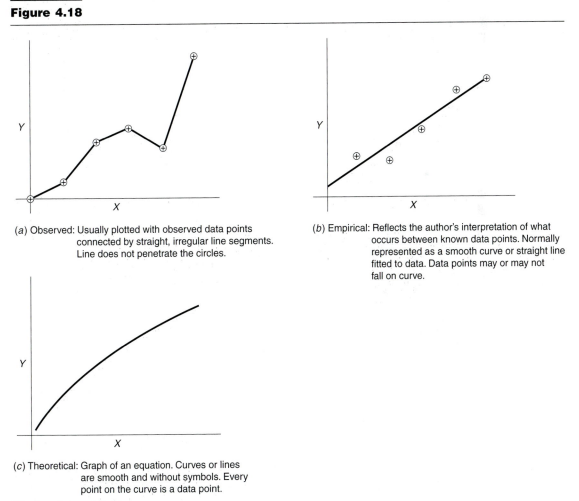

(*a*) Observed: Usually plotted with observed data points connected by straight, irregular line segments. Line does not penetrate the circles.

(*b*) Empirical: Reflects the author's interpretation of what occurs between known data points. Normally represented as a smooth curve or straight line fitted to data. Data points may or may not fall on curve.

(*c*) Theoretical: Graph of an equation. Curves or lines are smooth and without symbols. Every point on the curve is a data point.

Plotting data points.

Figure 4.19

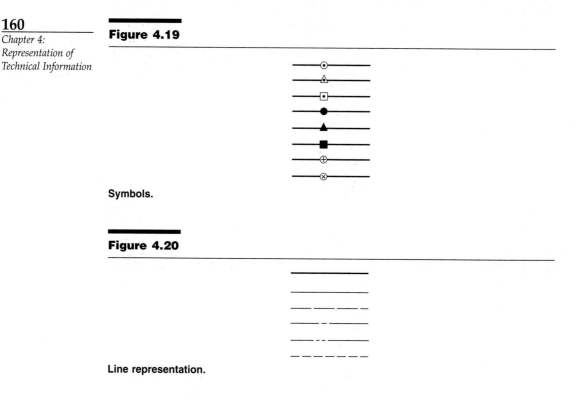

Symbols.

Figure 4.20

Line representation.

one curve is plotted on the same grid, several of these symbols may be used (one type for each curve). To avoid confusion, however, it is good practice to label each curve. When several curves are plotted on the same grid, another way they can be distinguished from each other is by using different types of lines, as illustrated in Fig. 4.20. Solid lines are normally reserved for single curves, and dashed lines are commonly used for extensions; however, different line representation can be used for each separate curve. The line weight of curves should be heavier than the grid ruling.

A key, or legend, should be placed in an available portion of the grid, preferably enclosed in a border, to define point symbols or line types that are used for curves. Remember, the lines representing each curve should never be drawn through the symbols so that the precise point is always identifiable. Figure 4.21 demonstrates the use of a key and the practice of breaking the line at each symbol.

4.3.7 Titles

Each graph must be identified with a complete title. The title should include a clear, concise statement of the data being represented, along with items such as the name of the author, the data of the experiment, and any and all information concerning the plot, including the name of the institution or company. Titles are normally enclosed in a border.

All lettering, the axes, and the curves should be sufficiently bold to stand out on the graph paper. Letters should be neat and of standard size. Figure 4.22

Figure 4.21

161

*General Graphing
Procedures*

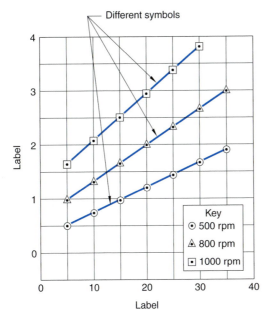

An example of a key.

is an illustration of plotted experimental data incorporating many of the items discussed in the chapter.

4.3.8 Computer-Assisted Plotting

A number of commercial software packages are available to produce graphs. The quality and accuracy of these computer-generated graphs vary depending on the sophistication of the software as well as on the plotter or printer employed. Typically the software will produce an axis scale graduated and calibrated to accommodate the range of data values that will fit the paper. This may or may not produce a readable or interpretable scale. Therefore it is necessary to apply considerable judgment depending on the results needed. For example, if the default plot does not meet needed scale readability, it may be necessary to specify the scale range to achieve an appropriate scale graduation since this option allows greater control of the scale drawn.

Computer-produced graphics with uniform scales may not follow the 1, 2, 5 rule, particularly since the software plots the independent variable based on the data collected. If the software has the option of separately specifying the range, that is, plotting the data as an X-Y scatter plot, you will be able to achieve scale graduations and calibrations that do follow the 1, 2, 5 rule, making it easier to read values from the graph. The hand-plotted graph that was illustrated in Fig. 4.22 is plotted using EXCEL. The results are shown in Fig. 4.23 using the X-Y scatter plot with a linear curve fit and the equation of the line using the method of least squares (see Sec. 4.6).

Figure 4.22

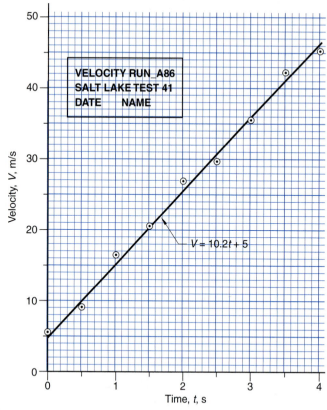

VELOCITY RUN_A86
SALT LAKE TEST 41
DATE   NAME

$V = 10.2t + 5$

Necessary steps to follow when manually plotting a graph:
1. Select the type of graph paper (rectilinear, semilog, log-log) and grid spacing for best representation of the given data.
2. Choose the proper location of the horizontal and vertical axes.
3. Determine the scale units (range) for each axis to display the data appropriately.
4. Graduate and calibrate the axes using the 1, 2, 5 rule.
5. Identify each axis completely.
6. Plot points using permissible symbols.
7. Check any point that deviates from the slope or curvature of the line.
8. Draw the curve or curves.
9. Identify each curve, add title and necessary notes.
10. Darken lines for good reproduction.

Sample plot.

Figure 4.23

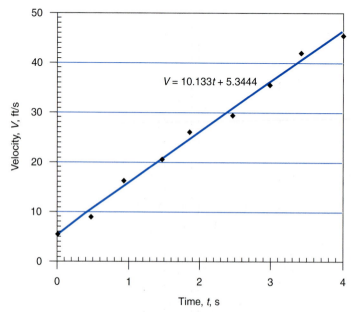

$V = 10.133t + 5.3444$

Necessary steps to follow when using a computer-assisted alternative:
1. Record via keyboard or import data into spreadsheet.
2. Select independent (x-axis) and dependent variable(s).
3. Select appropriate graph (style or type) from menu.
4. Produce trial plot with default parameters.
5. Examine (modify as necessary) origin, range, graduation, and calibrations: Note, use the 1, 2, 5 rule.
6. Label each axis completely.
7. Select appropriate plotting-point symbols and legend.
8. Create complete title.
9. Examine plot and store the data.
10. Plot or print the data.

EXCEL spreadsheet hard copy of data displayed in Fig. 4.22.

Empirical functions are generally described as those based on values obtained by experimentation. Since they are arrived at experimentally, equations normally available from theoretical derivations are not always possible. However, mathematical expressions can be modeled to fit experimental functions, and it is possible to classify most empirical results into one of four general categories: (1) linear, (2) exponential, (3) power, or (4) periodic.

A linear function, as the name suggests, will plot as a straight line on uniform rectangular coordinate paper. Likewise, when a curve representing experimental data is a straight line or a close approximation to a straight line, the relationship of the variables can be expressed by a linear equation, such as $y = mx + b$.

Correspondingly, exponential functions, when plotted on semilog paper, will be linear. How can that be? Since the basic form of the equation is $y = be^{mx}$, all we do is take the log of both sides. If it is written in log (base 10) form, it becomes $\log y = mx \log e + \log b$. Alternatively, using natural logarithms, the equation becomes $\ln y = mx + \ln b$ since $\ln e = 1$. The independent variable x is plotted on the abscissa and the dependent variable y is plotted on the functional $\ln$ (natural log) scale as $\ln y$.

The power equation has the form of $y = bx^m$. Written in log form it becomes $\log y = m \log x + \log b$. These data will plot as a straight line on log-log paper since the log of the independent variable x is plotted against the log of y.

When the data represent experimental results and a series of points are plotted to represent the relationship between the variables, it is improbable that a straight line can be constructed through every point since some error (instruments, readings, recordings) is inevitable. If all points do not lie on a straight line, an approximation scheme or averaging method may be used to arrive at the best possible fit. This method of straight-line approximation is called curve fitting.

4.5 Curve Fitting

Different methods or techniques are available to arrive at the best "straight-line" fit. Three methods commonly employed for finding the best fit are as follows:

1. Method of selected points
2. Method of averages
3. Method of least squares

Each of these techniques is progressively more accurate. The most accurate method, least squares, is discussed in considerably more detail in the chapter on statistics. However, several examples will be presented in this chapter to demonstrate correct methods for the representation of technical data using both methods of selected points and method of least squares.

Method 2, the method of averages, is based on the idea that the line location is positioned to make the algebraic sum of the absolute values of the differences between observed and calculated values of the ordinate equal to zero.

In both methods 2 and 3 the procedure involves minimizing what are called *residuals*, or the difference between an observed ordinate and the corresponding computed ordinate. The method of averages will not be covered in this book, but there are any number of reference texts available that adequately cover the concept.

4.6 Method of Selected Points and Least Squares

The method of selected points is a valid method of determining the equation that best fits data that exhibit a linear relationship. Once the data have been plotted and determined to be linear, a line is positioned that appears to fit the data best. This is accomplished most often by visually selecting a line that goes through as many data points as possible and has approximately the same number of data points on either side of the line.

Once the line has been constructed, two points, such as A and B, are selected *on the line* and at a reasonable distance apart. The coordinates of both points $A(X_1, Y_1)$ and $B(X_2, Y_2)$ must satisfy the equation of the line since both are points on the line.

The method of least squares is a much more accurate approach that will be illustrated as computer-assisted examples in most of the problems that follow. The method of least squares is the most appropriate technique for determination of the best-fit line. You should understand that the method presented represents a technique called *linear regression* and is valid only for *linear* relationships. The technique of least squares, however, can be applied to power ($y = bx^m$) and exponential ($y = be^{mx}$) relationships as well as linear ($y = mx + b$), if done correctly. The power function can be handled by noting that there is a linear relationship between log y and log x (log $y = m$ log x + log b, which plots as a straight line on log-log paper). Thus we can apply the method of least squares to the variables log y and log x to obtain parameters m and log b.

The exponential function written in natural logarithm form is ln $y = mx +$ ln b. Therefore a linear relationship exists between ln y and x (this plots as a straight line on semilog paper). The following sections will demonstrate the use of the selected point method and the computer-assisted method.

4.7 Empirical Equations—Linear

When experimental data plot as a straight line on rectangular grid paper, the equation of the line belongs to a family of curves whose basic equation is given by

$$y = mx + b \tag{4.1}$$

where m is the slope of the line, a constant, and b is a constant referred to as the *y intercept* (the value of y when $x = 0$).

To demonstrate how the method of selected points works, consider the following example.

Example problem 4.1 The velocity V of an experimental automobile is measured at specified time t intervals. Determine the equation of a straight

Table 4.3

Time *t*, s	0	5	10	15	20	25	30	35	40
Velocity *V*, m/s	24	33	62	77	105	123	151	170	188

line constructed through the points recorded in Table 4.3. Once an analytic equation has been determined, velocities at intermediate values can be computed.

Procedure

1. Plot the data on rectangular paper. If the results form a straight line (see Fig. 4.24), the function is linear and the general equation is of the form

    $$V = mt + b$$

    where *m* and *b* are constants.
2. Select two points on the line, $A(t_1, V_1)$ and $B(t_2, V_2)$, and record the value of these points. Points *A* and *B* should be widely separated to reduce the effect on *m* and *b* of errors in reading values from the graph. Points *A*

Figure 4.24

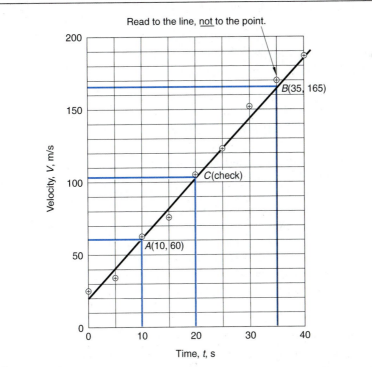

Data plot.

and B are identified on Fig. 4.24 for instructional reasons. They should not be shown on a completed graph that is to be displayed.

$$A(10, 60)$$

$$B(35, 165)$$

3. Substitute the points A and B into $V = mt + b$.

Eq(1) $60 = m(10) + b$

Eq(2) $165 = m(35) + b$

4. The equations are solved simultaneously for the two unknowns.

$$m = 4.2$$

$$b = 18$$

5. The general equation of the line for this specific problem can be written as

$$V = 4.2t + 18$$

6. Using another point $C(t_3, V_3)$, check for verification.

$$C(20, 102)$$

$$102 = 4.2(20) + 18$$

$$102 = 84 + 18 = 102$$

7. *A computer-assisted alternative.*   The data set can be entered into a spreadsheet package and the software will provide a more precise solution. Figure 4.23 illustrates a software solution for data provided in the previous sample plot (Fig. 4.22). The computer solution provides a plot of data on rectilinear paper with the equation of the line determined by the method of least squares. Once you understand the fundamentals, it can be very time efficient to use computer technology and commerical software.

4.8 Empirical Equations—Power Curves

When experimentally collected data are plotted on rectangular coordinate graph paper and the points do not form a straight line, you must determine which family of curves the line most closely approximates. If you have no idea as to the nature of the data, plot the experimentally collected points on log-log paper or semilog paper to determine if the data approximate a straight line. Consider the following familiar example. Suppose that a solid object is dropped from a tall building. To anyone who has studied fundamental physics, it is apparent that these values should correspond to the general equation for a free-falling body (neglecting air friction):

$$s = 1/2gt^2$$

However, let us assume for a moment that we do not know this relationship and that all we have is a table of values experimentally measured on a free-falling body.

Table 4.4

Time t, s	Distance s, m
0	0
1	4.9
2	19.6
3	44.1
4	78.4
5	122.5
6	176.4

Example problem 4.2   A solid object is dropped from a tall building, and the values time versus distance are as recorded in Table 4.4.

Procedure

1. Make a freehand plot to observe the data visually (see Fig. 4.25). From this quick plot the data points are more easily recognized as belonging to a family of curves whose general equation can be written as

$$y = bx^m \tag{4.2}$$

Remember, before the method of selected points can be applied to determine the equation of the line, the plotted line must be straight because two points on a curved line do not uniquely identify the line. Mathematically, this general equation can be modified by taking the logarithm of both sides,

$$\log y = m \log x + \log b \quad \text{or} \quad \ln y = m \ln x + \ln b$$

This equation suggests that if the logs of all table values of y and x were determined and the results were plotted on rectangular paper, the line would likely be straight.

Realizing that the log of zero is undefined and plotting the remaining points that are recorded in Table 4.5 for log s versus log t, the results are shown in Fig. 4.26.

Table 4.5

Time t, s	Distance s, m	Log t	Log s
0	0		
1	4.9	0.0000	0.6902
2	19.6	0.3010	1.2923
3	44.1	0.4771	1.6444
4	78.4	0.6021	1.8943
5	122.5	0.6990	2.0881
6	176.4	0.7782	2.2465

Figure 4.25

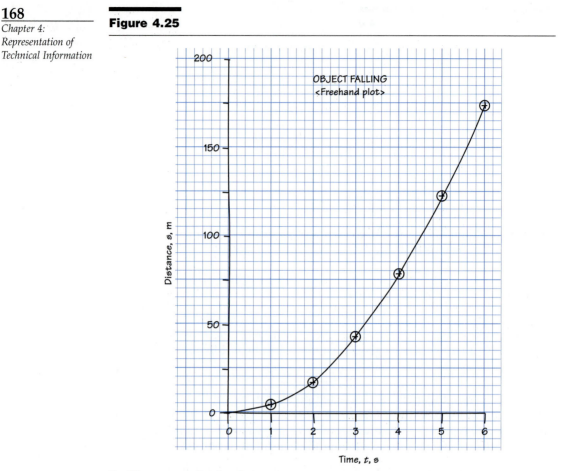

Rectilinear paper (freehand).

Since the graph of log s versus log t does plot as a straight line, it is now possible to use the general form of the equation

$$\log y = m \log x + \log b$$

and apply the method of selected points.

When reading values for points A and B from the graph, we must remember that the logarithm of each variable has already been determined and the values have been plotted.

$$A(0.2, 1.09)$$

$$B(0.6, 1.89)$$

Points A and B can now be substituted into the general equation log $s = m$ log $t +$ log b and solved simultaneously.

$$1.89 = m(0.6) + \log b$$

$$1.09 = m(0.2) + \log b$$

$$m = 2.0$$

Figure 4.26

169

*Empirical
Equations—
Power Curves*

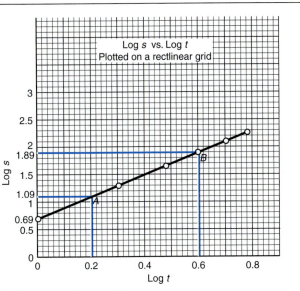

Plotting log of variables on rectilinear grid paper.

$$\log b = 0.69 \qquad \text{or} \qquad b = 4.9$$

An examination of Fig. 4.26 shows the value of $\log b$ (0.69) can be read from the graph where $\log t = 0$. This, of course, is where $t = 1$ and is the y intercept for log-log plots.

The general equation can then be written as

$$s = 4.9t^{2.0} \qquad \text{or} \qquad s = 1/2gt^2,$$

where $g = 9.8 \text{ m/s}^2$.

Note: One obvious inconvenience is the necessity of finding logarithms of each variable and then plotting the logs of these variables. This step is not necessary since functional paper is commercially available with $\log x$ and $\log y$ scales already constructed. Log-log paper allows the variables themselves to be plotted directly without the need of computing the log of each value.

2. An alternate method for the solution of this problem follows.

   In the preceding example, once the general form of the equation is determined [Eq. (4.2)], the data can be plotted directly on log-log paper. Since the resulting curve is a straight line, the method of selected points can be used directly (see Fig. 4.27).

   The log form of the equation is again used.

$$\log s = m \log t + \log b$$

Select points A and B on the line.

$$A(1.5, 11)$$

$$B(6, 175)$$

Figure 4.27

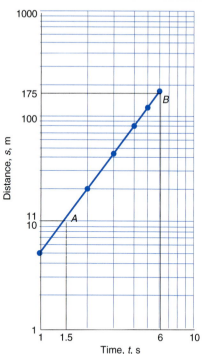

Log-log paper.

Substitute the values into the general equation $\log s = m \log t + \log b$, taking careful note that the numbers are the variables and *not* the logs of the variables.

$$\log 175 = m \log 6 + \log b$$

$$\log 11 = m \log 1.5 + \log b$$

Again, solving these two equations simultaneously results in the following approximate values for the constants b and m:

$$b = 4.8978 = 4.9$$

$$m = 1.9957 = 2.0$$

Identical conclusions can be reached:

$$s = 1/2gt^2$$

This time, however one can use functional scales rather than calculate the log of each number.

3. *A computer-assisted alternative.* The data set can be entered into a spreadsheet package and the software will provide an identical solution. Figure 4.28 illustrates software that provides a plot of data on rectilinear paper including the equation of the line. Figure 4.29 is an example of the data plotted on log-log paper with the software providing the equation.

Figure 4.28

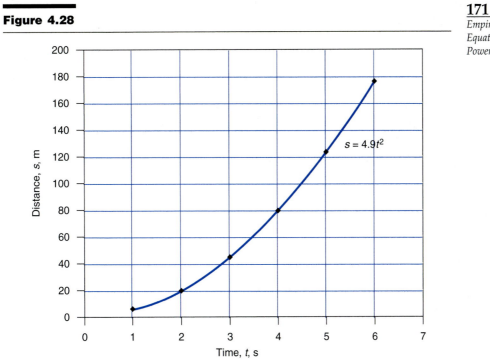

Free-falling object on rectilinear paper.

Figure 4.29

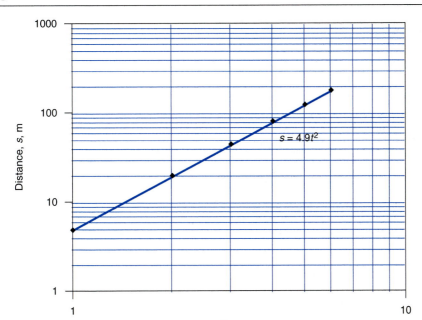

Free-falling object on log-log paper.

4.9 Empirical Equations—Exponential Curves

Suppose that your data do not plot as a straight line on rectangular coordinate paper nor is the line approximately straight on log-log paper. Without experience in analyzing experimental data, you may feel lost as to how to proceed. Normally when experiments are conducted, you have an idea as to how the parameters are related and are merely trying to quantify that relationship. If you plot your data on semilog graph paper and it produces a reasonably straight line, then it has the general form

$$y = be^{mx} \tag{4.3}$$

Example problem 4.3   A rocket sled fuel consumption is recorded as shown in Table 4.6.

Procedure

1. The data (Table 4.6) when plotted produce the graph shown in Fig. 4.30. To determine the constants in the equation $y = be^{mx}$, write it in linear form, as either

$$\log y = mx \log e + \log b \quad \text{or} \quad \ln y = mx + \ln b$$

   The method of selected points can now be employed for $\ln FC = mV + \ln b$ (choosing the natural log form). Points $A(15, 33)$ and $B(65, 470)$ are carefully selected on the line, so that they must satisfy the equation. Substituting the values of V and FC at points A and B, we get

$$\ln 470 = 65\, m + \ln b \quad \text{and} \quad \ln 33 = 15\, m + \ln b$$

   Solving simultaneously for m and b, we have

$$m = 0.0529 \quad \text{and} \quad b = 14.8$$

   The desired equation then is determined to be $FC = 15e^{(0.05V)}$. This determination can be checked by choosing a third point, substituting the value for V, and solving for FC.

2. *A computer-assisted alternative.*   The data set can be entered into a spreadsheet package and the software will provide an identical solution. Figure 4.31 illustrates software that provides a plot of data on rectilinear

---

Table 4.6

Velocity V, m/s	Fuel Consumption (FO), mm³/s
10	25.2
20	44.6
30	71.7
40	115
50	202
60	367
70	608

Figure 4.30

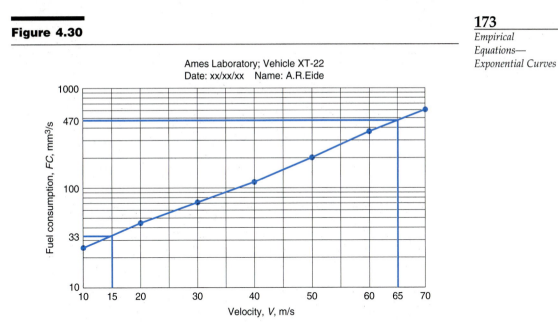

Ames Laboratory; Vehicle XT-22
Date: xx/xx/xx   Name: A.R.Eide

Semilog paper.

Figure 4.31

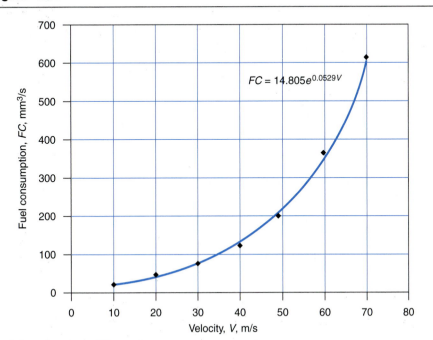

$$FC = 14.805e^{0.0529V}$$

Rocket engine on rectilinear paper.

Figure 4.32

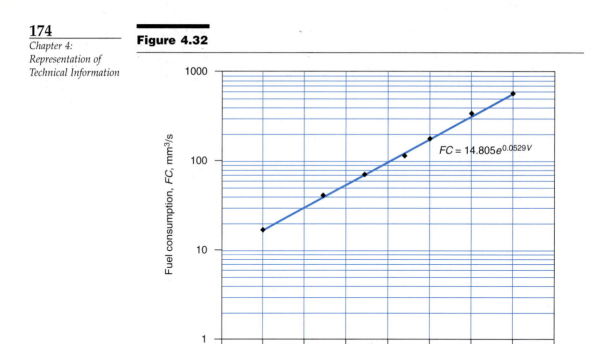

Rocket engine on semilog paper.

paper including the equation of the line. Figure 4.32 is an example of the data plotted on semilog paper with the software providing the equation.

Key Terms and Concepts

The following are some of the terms and concepts you should recognize and understand.

Scale graduations	**Line representation**
Scale calibrations	**Titles**
Uniform scales	**Empirical functions**
Nonuniform scale	**Curve fitting**
1, 2, 5 rule	**Method of selected points**
Plotting procedures	**Linear curves**
Graph symbols	**Power curves**
Axis identification	**Exponential curves**

Problems

Plotting Problems

4.1 Table 4.7 shows data on the CO_2 levels in the atmosphere from 1960 to 1990. Plot the data on rectilinear paper using proper plotting procedures.

4.2 Imagine that you received a $5000.00 gift for high school graduation. Instead of making a down payment on a new car, you decided to invest the entire amount

Table 4.7

Year	CO$_2$ levels, ppm	Year	CO$_2$ levels, ppm
1960	316	1976	331
1961	316.25	1977	334
1962	318	1978	335
1963	319	1979	335.7
1964	319.8	1980	339
1965	319.5	1981	339.9
1966	320.5	1982	340
1967	321	1983	341.5
1968	323	1984	344.6
1969	324.25	1985	345
1970	324.8	1986	345.5
1971	325.5	1987	349
1972	326	1988	350.8
1973	329.7	1989	352
1974	330	1990	353
1975	330.5		

in a market fund that paid annual interest, compounded monthly. Table 4.8 illustrates the return if the money yields 5, 10, and 15 percent interest. On a single sheet of rectilinear paper, plot the 10-year returns.

4.3 Bacterial growth can be estimated by observation. A new infection has the following recorded rate of growth in Table 4.9:

(*a*) Plot the data on rectilinear paper using proper plotting procedures.

(*b*) Plot the data on semilog paper using proper plotting procedures.

4.4 The number of events (migrating birds) observed each year were recorded in Table 4.10.

(*a*) Plot the data on rectilinear paper using proper plotting procedures.

(*b*) Plot the data on log-log paper using proper plotting procedures.

Method of Selected Points

4.5 Table 4.11 contains data from a constant acceleration trial run by an experimental electric-powered vehicle recording velocity against time.

(*a*) Plot the data on rectilinear paper using time as the independent variable.

Table 4.8

Year	Return @ 5% Dollars	Return @ 10% Dollars	Return @ 15% Dollars
1	5256	5524	5804
2	5525	6102	6737
3	5807	6741	7820
4	6105	7447	9077
5	6417	8227	10536
6	6745	9088	12230
7	7090	10040	14197
8	7453	11091	16478
9	7834	12252	19126
10	8235	13535	22201

Table 4.9

Hour	Rate of Growth Cells/hour
1	16.5
2	27.2
3	44.8
4	73.9
5	121.8
6	200.8
7	331
8	545
9	900
10	1484

Table 4.10

Year	Number of Sightings × 10³
1985	2.5
1986	9.3
1987	20.2
1988	34.8
1989	53.3
1990	75.2
1991	100.8
1992	130
1993	162
1994	198.6
1995	238
1996	280
1997	327
1998	376
1999	430

Table 4.11

Time, t, s	Velocity, V, ft/s
0	0
5	65
10	120
15	175
20	240
25	307
30	370
35	440
40	515
45	580
50	650
55	720
60	790

    (b) Determine the equation of the line using the method of selected points.
    (c) Interpret the slope of the line.
4.6 Table 4.12 lists the values of velocity versus time as a skateboard rider inclines down a 45° (degree) slope.
    (a) Plot the data on rectilinear paper using time as the independent variable.
    (b) Using the method of selected points, determine the equation of the line.
    (c) Estimate the average acceleration.
4.7 Table 4.13 shows values of temperatures in degrees Fahrenheit and Celsius.
    (a) Plot the data on a linear plot.
    (b) What is the temperature in Celsius for 84°F?
    (c) What is the temperature in Fahrenheit for 84°C?
4.8 Table 4.14 is a collection of data for an iron-constantan thermocouple. Temperature is in degrees Celsius and the electromotive force (emf) is in millivolts.
    (a) Plot a graph, using rectilinear paper, showing the relation of temperature to voltage, with voltage as the independent variable.
    (b) Using the method of selected points, find the equation of the line.
4.9 A new experimental pump was tested to determine the power required to produce a range of discharges. Various rates of discharge were measured and the

Table 4.12

Time, t, s	Velocity, V, m/s
0	0
10	14
20	29
30	44
40	62
50	78
60	95

Table 4.13

Celsius	Fahrenheit
−60	−76
−40	−40
−20	−4
0	32
20	68
60	140
80	176
100	212
120	248

Table 4.14

Voltage, emf, V	Temp, T, C
5	82
10	175
15	270
20	365
25	460
30	555
35	650
40	745
45	840
50	935

Table 4.15

Discharge, Q, L/s	Power, P, kW
5	31
10	39
15	45
20	53
25	60
30	67
35	75

corresponding power required for each discharge was recorded. The results of the test are shown in Table 4.15.
(a) Plot a graph showing the power required for various rates of discharge.
(b) Determine the equation of the relationship.
(c) Extrapolate the power required to produce a discharge of 40 L/s.

4.10 The area of a circle can be expressed by the formula $A = \pi R^2$. If the radius varies from 0.5 to 5 cm, perform the following:
(a) Construct a table of radius versus area mathematically. Use radius increments of 0.5 cm.
(b) Construct a second table of log R versus log A.
(c) Plot the values from (a) on log-log paper and determine the equation of the line.
(d) Plot the values from (b) on rectilinear paper and determine the equation of the line.

4.11 Electrical resistance for a given material can be a function of both area/unit thickness and material temperature. Holding temperature constant a range of areas are tested to determine resistance. The measured resistance recorded in Table 4.16 is expressed in milliohms per meter of conductor length.
(a) Plot data on rectilinear paper, with area as the independent variable.
(b) Plot the data on log-log graph paper, with area as the independent variable.
(c) Plot the data on semilog graph paper, with area as the independent variable.
(d) What would be the best curve fit for this application?

Table 4.16

Area, A, mm²	Resistance, R, mΩ/m
0.05	215
0.1	110
0.2	57
0.5	23
1	12
3	4
5	2.5
10	1.3

Table 4.17

Voltage, V, volts	Current, I, amps
1.5	0.29
3	0.61
5	1.06
9	1.78
12	2.42
15	3
20	4.2
30	5.9
50	9.8

Method of Selected Points/Computer Assisted

4.12 The voltage was changed across a 5 Ω (ohms) resistor and the current was measured using an ammeter. Values were recorded in Table 4.17.
   (a) Plot on rectilinear paper and determine the equation of the line using the method of selected points.
   (b) Using a computer-assisted method, plot the graph and determine the equation of the line.
4.13 A Lamborghini Diablo is able to reach a maximum speed of 208 mph (miles per hour). Table 4.18 shows the time in seconds that it takes for the Diablo to accelerate from 30 to 130 mph.
   (a) Plot on rectilinear paper and determine the equation of the line using the method of selected points.
   (b) Using a computer-assisted method, plot the graph and determine the equation of the line.
   (c) Compare the two methods and comment on any differences.
4.14 A new production facility manufactured 26 parts the first month, but then increased production, as shown in Table 4.19.
   (a) Plot the data on semilog paper.
   (b) Using the time variable that defines January to be 1, February to be 2, and so on, determine the equation that fits the data.

Table 4.18

Velocity, V, mph	Seconds
30	2.1
40	2.9
50	3.6
60	4.3
70	5.3
80	6.4
90	7.4
100	8.2
110	9.7
120	11.8
130	13.7

Table 4.19

Month	Number of Parts/Month
1	26
2	33
3	42
4	54
5	70
6	90
7	115
8	150
9	190
10	245
11	312
12	400

Table 4.20

Plate Thickness, inches	Geiger Reading, rad/s
0.1	5950
0.5	5780
1	5565
2.5	4975
5	4125
10	2835
20	1340
30	630
40	300
50	140

4.15 The rate of absorption of radiation by metal plates varies with the plate thickness and the nature of the source of radiation. A test was made at Ames Labs on October 11, 1982, using a Geiger counter and a constant source of radiation; the results are shown in Table 4.20.

(a) Plot the data on semilog graph paper.

(b) Find the equation of the relationship between the parameters.

(c) What level of radiation would you estimate to pass a 2 in thick plate of the metal used in the test described earlier?

4.16 The volume of a cone is $V = (\pi r^2 h)/3$. Assume that the height is constant at 3 cm (centimeters).

(a) Develop a table showing the volume and radius, as the radius changes from 1.0 to 12.0 cm in 1 cm increments.

(b) Plot the data on log-log paper and determine the equation of the relationship by the method of selected points.

(c) Using a computer-assisted method, plot the graph and determine the equation of the line.

4.17 As atmospheric pressure increases, water will begin boiling at higher temperatures. Table 4.21 provides experimental observations of this change in boiling temperature with increased pressure.

Table 4.21

Pressure, P, lbf/in^2	Temperature, °F
10	193.2
20	228
40	267.3
60	292.7
80	312
100	327.8
200	381.9
300	417.4
400	444.7
500	467
1000	544.7
2000	636
3000	695

Figure 4.33

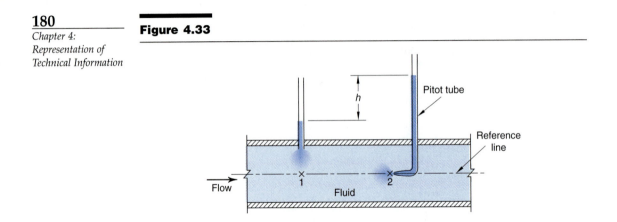

Pitot tube

Reference line

h

Flow

Fluid

Table 4.22

Height, h, m	Velocity, V, m/s
0.1	1.4
0.2	2
0.4	2.8
0.6	3.4
0.8	4
1	4.4

(a) Plot the data on log-log paper and determine the equation of the relationship by the method of selected points.

(b) Using a computer-assisted method, plot the graph and determine the equation of the line.

4.18 A pitot tube is a device for measuring the velocity of flow of a fluid (see Fig. 4.33). A stagnation point occurs at point 2; by recording the height differential h, the velocity at point 1 can be calculated. Assume for this problem that the velocity at point 1 is known corresponding to the height differential h. Table 4.22 records these values.

(a) Plot the data on log-log paper using height as the independent variable.

(b) Determine the equation of the line using the method of selected points.

Computer Assisted

4.19 The power function represents the surface area and volume of a certain geometric shape. Given the values in Table 4.23, determine that shape.

(a) Using computer-assisted software determine the equations of the two variables.

(b) What is the geometric shape?

4.20 A new start-up Internet company offered a free download of music to all web users during its first 30 days of operation. At the end of the third week they had provided downloads as shown in Table 4.24.

(a) Using computer-assisted software, determine the equations that best fits the data assuming that downloads continue exponentially.

(b) Determine the number of downloads that can be projected on day 30.

Table 4.23

Radius R, ft	Surface Area SA, ft^2	Volume V, ft^3
1	12.6	4.19
2	50.3	33.5
3	113	113
4	200	268
5	314	524
6	450	905
7	616	1437
8	800	2145
9	1018	3054
10	1257	4189

Table 4.24

Day	Number of Downloads
1	20
2	23
3	27
4	34
5	41
6	50
7	60
8	75
9	91
10	110
11	136
12	165
13	202
14	248
15	301
16	368
17	450
18	550
19	670
20	820
21	1000

4.21 Tests determined the heat produced by a furnace, in thousands of British thermal units (Btu) per cubic feet (ft^3) of furnace volume, at different temperatures. Table 4.25 records the results.
 (*a*) With computer-assisted software, plot the data using rectilinear scales.
 (*b*) With computer-assisted software, plot the data using a dependent semilog scale.
 (*c*) With computer-assisted software, plot the data using a log-log scale.
 (*d*) Determine the equation of the line in part (*c*).
4.22 From the information in Prob. 4.21, determine
 (*a*) amount of heat released at 550°F
 (*b*) amount of heat released at 1200°F (projected)
 (*c*) Temperature at 5000 Btu/ft^3
 (*d*) Temperature at 40000 Btu/ft^3 (projected)

Table 4.25

Temperature, T, F	Heat Released, HR, Btu/ft^3
100	30
200	240
300	810
400	1920
500	3750
600	6480
700	10290
800	15360
900	21870
1000	30000

CHAPTER 5

Engineering Estimations and Approximations

5.1 Introduction

Previously we discussed a myriad of engineering disciplines and functions. Within each discipline there are specialty areas which give the appearance that engineering is a diverse profession with little commonality in the tasks performed. However, engineers are problem solvers, creating new designs which satisfy a need and improve the living standard. During the design process engineers of all disciplines will need to acquire physical measurements pertaining to the product or system being designed, the environment in which the design will operate, or both.

The nineteenth-century physicist Lord Kelvin stated that knowledge and understanding are not of high quality unless the information can be expressed in numbers. We all have made or heard statements, such as "the water is too hot." This statement may or may not give us an indication of the temperature of the water. At a given temperature water may be too hot for taking a bath but not hot enough for making instant coffee or tea.

The truth is that pronouncements, such as "hot," "too hot," "not very hot," and so on, are relative to a standard selected by the speaker and have meaning only to those who know that standard.

Engineers make measurements of a vast array of physical quantities that control the design solution. Skill in making and interpreting measurements is an essential element in our practice of engineering.

5.2 Significant Digits

Any physical measurement cannot be assumed to be exact. Errors are likely to be present regardless of the precautions used when making the measurement. Quantities determined by analytical means are not always exact either. Often assumptions are made to arrive at an analytical expression, which is then used to calculate a numerical value.

It is clear that a method of expressing results and measurements is needed that will convey how "good" these numbers are. The use of significant digits gives us this capability without resorting to the more rigorous approach of

Figure 5.1

Quantity	Number of Significant Figures
4784	4
36	2
60	1 or 2
600	1, 2, or 3
6.00×10^2	3
31.72	4
30.02	4
46.0	3
0.02	1
0.020	2
600.00	5

computing an estimated percentage error to be specified with each numerical result or measurement.

A *significant digit,* or *figure,* is defined as any digit used in writing a number, *except* those zeros that are used only for location of the decimal point or those zeros that do not have any nonzero digit on their left. When you read the number 0.0015, only the digits 1 and 5 are significant since the three zeros have no nonzero digit to their left. We would say then that this number has two significant figures. If the number is written 0.00150, it contains three significant figures; the rightmost zero is significant.

Numbers 10 or larger that are not written in scientific notation and that are not counts (exact values) can cause difficulties in interpretation when zeros are present. For example, 2000 could contain one, two, three, or four significant digits; it is not clear which. If you write the number in scientific notation as 2.000×10^3, then clearly four significant digits are intended. If you want to show only two significant digits, you would write 2.0×10^3. It is our recommendation that, if uncertainty results from using standard decimal notation, you switch to scientific notation so that your reader can clearly understand your intent. Figure 5.1 shows the number of significant figures for several quantities.

You may find yourself as the user of values where the writer was not careful to show significant figures properly. What then? Assuming that the number is not a count or a known exact value, about all you can do is establish a reasonable number of significant figures based on the context of the value and on your experience. Once you have decided on a reasonable number of significant digits, you can then use the number in any calculations that are required.

When reading instruments, such as an engineer's scale, analog thermometer, or fuel gage, the last digit will normally be an estimate. That is, the instrument is read by estimating between the smallest graduations on the scale to get the final digit. In Fig. 5.2a you may estimate the reading from the engineer's scale to be 1.27, with the 7 being a doubtful digit of the three significant figures.

Figure 5.2

185

Significant Digits

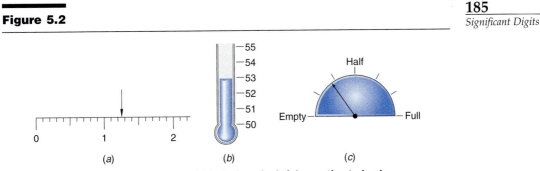

(a)   (b)   (c)

Reading graduations on instruments will include a doubtful or estimated value.

It is standard practice to count one doubtful digit as significant, thus the 1.27 reading has three significant figures. Similarly, the thermometer in Fig. 5.2b may be read as 52.9° with the 9 being doubtful.

In Fig. 5.2c the graduations create a more difficult task for reading a fuel level. Each graduation is one-sixth of a full tank. The reading appears to be about three-fourths of the distance between one-sixth and two-sixths making the reading seven twenty-fourths of a tank or 0.292. How many significant figures are there? In this case one significant figure is all that can be obtained, so the answer should be rounded to 0.3. The difficulty in this example is not the significant figures, but the scale of the fuel gage. It is meant to convey a general impression of the fuel level and not a numerically significant value. Furthermore, the automobile manufacturer did not deem that the cost of a more accurate and precise fuel measurement system was justified. Therefore the selection of the instrument is an important factor in physical measurements.

Calculators and computers commonly work with numbers having as few as 7 digits or as many as 16 or 17 digits. This is true no matter how many significant digits an input value or calculated value should have. Therefore you will need to exercise care in reporting values from a calculator display or from a computer output (a spreadsheet, e.g.). Some calculators and most high-level computer languages allow you to control the number of digits that are to be displayed or printed. If a computer output is to be a part of your final solution presentation, you will need to control the output form carefully. If the output is only an intermediate step, you can round the results to a reasonable number of significant figures in your presentation.

As you perform arithmetic operations, it is important that you do not lose the significance of your measurements or, conversely, imply precision that does not exist. Rules for determining the number of significant figures that should be reported following computations have been developed by engineering associations. The following rules customarily apply:

1. *Multiplication and division.* The product or quotient should contain the same number of significant digits as those contained in the number with the fewest significant digits.

    Examples
    a. $(2.43)(17.675) = 42.95025$

If each number in the product is exact, the answer should be reported as 42.95025. If the numbers are not exact, as is normally the case, 2.43 has three significant figures and 17.675 has five. Applying the rule, the answer should contain three significant figures and should be reported as 43.0 or 4.30×10^1.

b. $(2.479 \text{ h})(60 \text{ min/h}) = 148.74 \text{ min}$

In this case the conversion factor is exact (a definition) and could be thought of as having an infinite number of significant figures. Thus 2.479, which has four significant figures, controls the precision, and the answer is 148.7 min, or 1.487×10^2 min.

c. $(4.00 \times 10^2 \text{ kg})(2.2046 \text{ lbm/kg}) = 881.84 \text{ lbm}$

Here the conversion factor is not exact, but you should not let the conversion factor dictate the precision of the answer if it can be avoided. You should attempt to maintain the precision of the value being converted; you cannot improve its precision. Therefore you should use a conversion factor that has one or two more significant figures than will be reported in the answer. In this situation three significant figures should be reported, yielding 882 lbm (pound-mass).

d. $589.62/1.246 = 473.21027$

The answer, to four significant figures, is 473.2.

2. *Addition and subtraction.* The answer should show significant digits only as far to the right as is seen in the least precise number in the calculation. Remember, the last number recorded is doubtful, that is, an estimate.

Example

a.
```
  1725.463
   189.2
    16.73
  1931.393
```

The least precise number in this group is 189.2 because the (0.2) is an estimate, so, according to the rule, the answer should be reported as 1931.4. Using alternative reasoning, suppose that these numbers are instrument readings, which means the last reported digit in each is a doubtful digit. A column addition that contains a doubtful digit will result in a doubtful digit in the sum. So all three digits to the right of the decimal in the answer are doubtful. Normally we report only one; thus the answer is 1931.4 after rounding.

b.
```
   897.0
    -0.0922
   896.9078
```

Application of the rule results in an answer of 896.9.

3. *Combined operations.* If products or quotients are to be added or subtracted, perform the multiplication or division first, establish the correct number of significant figures in the subanswer, perform the addition or subtraction and round to proper significant figures. Note, however, that in calculator or in computer applications it is not practical to perform intermediate rounding. It is normal practice to perform the entire calculation and then report a reasonable number of significant figures.

If results from additions or subtractions are to be multiplied or divided, an intermediate determination of significant figures can be made when the calculations are performed manually. Use the suggestion already mentioned for calculator or computer answers.

Subtractions that occur in the denominator of a quotient can be a particular problem when the numbers to be subtracted are very nearly the same. For example, 39.7/(772.3 − 772.26) gives 992.5 if intermediate roundoff is not done. If, however, the subtraction in the denominator is reported with one digit to the right of the decimal, the denominator becomes zero and the result becomes undefined. Commonsense application of the rules is necessary to avoid problems.

4. *Rounding.* In rounding a value to the proper number of significant figures, *increase the last digit retained by 1 if the first figure dropped is 5 or greater.* This is the rule normally built into a calculator display control or a control language.

Examples

a.   827.48 rounds to 827.5 or 827 for four and three significant digits, respectively.

b.   23.650 rounds to 23.7 for three significant figures.

c.   0.0143 rounds to 0.014 for two significant figures.

5.3   Accuracy and Precision

In measurements accuracy and precision have different meanings and cannot be used interchangeably. *Accuracy* is a measure of the nearness of a value to the correct or true value. *Precision* refers to the repeatability of a measurement, that is, how close successive measurements are to each other. Figure 5.3 illustrates accuracy and precision of the results of four dart throwers. Thrower (*a*) is both inaccurate and imprecise because the results are away from the bull's-eye (accuracy) and widely scattered (precision). Thrower (*b*) is accurate because the throws are evenly distributed about the desired result but imprecise because of the wide scatter. Thrower (*c*) is precise with the tight cluster of throws but inaccurate because the results are away from the desired bull's-eye. Finally, thrower (*d*) demonstrates accuracy and precision with a tight cluster of throws around the center of the target. Throwers (*a*), (*b*), and (*c*) can improve their performance by analyzing the causes for the errors. Body position, arm motion, and release point could cause deviation from the desired result.

Engineers making physical measurements encounter two types of errors; systematic and random. These will be discussed in the next section.

Measurements can be reported as a value plus or minus (±) a number, for example, 32.3 ± 0.2. This indicates a range of values which are equally representative of the indicated value (32.3). Thus 32.3, 32.1, and 32.5 are among the "acceptable" values for this measurement. A range of permissible error also can be specified as a percentage of the indicated value. For example, a thermometer's accuracy may be specified as ±1.0 percent of full-scale reading. Thus if the full-scale reading is 220°F, readings should be within ±2.2 of the true value (220 × 0.01 = 2.2).

Figure 5.3

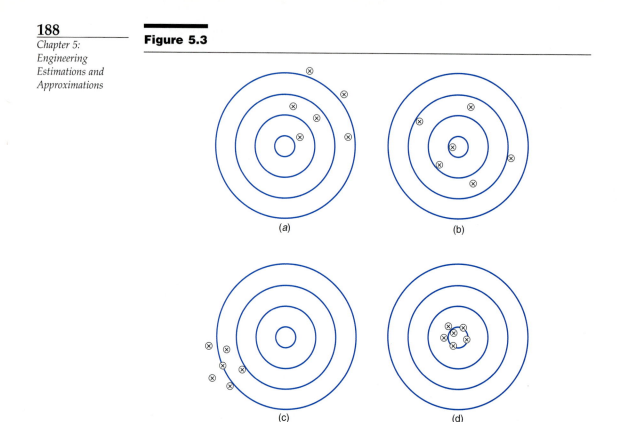

(a)　　　　　(b)

(c)　　　　　(d)

Illustration of the difference between accuracy and precision in physical measurements.

5.4   Errors

To measure is to err! Any time a measurement is taken, the result is being compared to a true value, which itself may not be known exactly. If we measure the dimensions of a room, why does a repeat of the measurements not yield the same results? Did the same person make all measurements? Was the same measuring instrument used? Were the readings all made from exactly the same eye position? Was the measuring instrument correctly graduated? It is obvious that errors will occur in each measurement. We must try to identify the errors if we can and to correct them in our results. If we cannot identify the error, we must provide some conclusions as to the resulting accuracy and precision of our measurements.

Identifiable and correctable errors are classified as systematic; accidental or other nonidentifiable errors are classified as random.

5.4.1   Systematic Errors

Our task is to measure the distance between two fixed points. Assume that the distance is about 1200 m and that we are experienced and competent and have equipment of high quality to do the measurement. Some of the errors that occur will always have the same sign (+ or −) and are said to be systematic. Assume that a 25 m steel tape is to be used, one that has been compared with the standard at the U.S. Bureau of Standards in Washington, DC. If the tape is not exactly 25.000 m long, then there will be a systematic error each of the 48 times that we use the tape to measure out the 1200 m.

However, the error can be removed by applying a correction. A second source of error can stem from a difference between the temperature at the time of use and when the tape was compared with the standard. Such an error can be removed if we measure the temperature of the tape and apply a mathematical correction. The coefficient of thermal expansion for steel is 11.7×1.0^{-6} per kelvin. The accuracy of such a correction depends on the accuracy of the thermometer and on our ability to measure the temperature of the tape instead of the temperature of the surrounding air. Another source of systematic error can be found in the difference in the tension applied to the tape while in use and the tension employed during standardization. Again, scales can be used but, as before, their accuracy will be suspect. In all probability the tape was standardized by laying it on a smooth surface and supporting it throughout. Such surfaces, however, are seldom available in the field. The tape is suspended at times, at least partially. But knowing the weight of the tape, the tension that is applied, and the length of the suspended tape, we can calculate a correction and apply it.

The sources of systematic error just discussed are not all the possible sources, but they illustrate an important problem encountered even in taking comparatively simple measurements. Similar problems occur in all types of measurements: mechanical quantities, electrical quantities, mass, sound, odors, and so forth. We must be aware of the presence of systematic errors, eliminate those that we can, and quantify and correct for those remaining.

5.4.2 Random Errors

In reading the previous section, you may have realized that even if it had been possible to eliminate all the systematic errors, the measurement is still not exact. To elaborate on this point, we will continue with the example of the task of measuring the 1200 m distance. Several random errors can creep in, as follows. When reading the thermometer, we must estimate the reading when the indicator falls between graduations. Moreover, it may appear that the reading is exactly on a graduation when it is actually slightly above or below the graduation. Furthermore, the thermometer may not be accurately measuring the tape temperature but, may be influenced instead, by the temperature of the ambient air. These errors thus can produce measurements that are either too large or too small. Regarding sign and magnitude, the error therefore is random.

Errors also can result from our correcting for the sag in a suspended tape. In such a correction it is necessary to determine the weight of the tape, its cross-sectional area, its modulus of elasticity, and the applied tension. In all such cases the construction of the instruments used for acquiring these quantities can be a source of both systematic and random errors.

The major difficulty we encounter with respect to random errors is that, although their presence is obvious by the scatter in the data, it is impossible to predict the magnitude and sign of the accidental error that is present in any one measurement. Repeating measurements and averaging the results will reduce the random error on the average. However, repeating measurements will not reduce the systematic error in the average result.

Refinement of the apparatus and care in its use can reduce the magnitude of the error; indeed, many engineers have devoted their careers to this task.

Likewise, awareness of the problem, knowledge about the degree of precision of the equipment, skill with measurement procedures, and proficiency in the use of statistics allow us to estimate the magnitude of the error remaining in measurements. This knowledge in turn allows us to accept the error or to develop different apparatus and/or methods in our work. It is beyond the scope of this text to discuss quantifying accidental errors. However, Chapter 8 includes a brief discussion of central tendency and standard deviation, which are part of the analysis of random errors.

5.5 Approximations

Engineers strive for high-level precision in their work. However, it is also important to be aware of an acceptable precision and the time and cost of attaining it. There are many instances where an engineer is expected to make an approximation to an answer, that is, to estimate the result with reasonable accuracy but under tight time and cost constraints. To do this, engineers rely on their basic understanding of the problem under discussion coupled with their previous experience. This knowledge and experience is what distinguishes an "approximation" from a "guess." If greater accuracy is needed, the initial approximation can be refined when time and funds are available and the necessary additional data for refining the result are available.

In the area of our highest competency we are expected to be able to make rough estimates to provide figures that can be used for tentative decisions. These estimates may be in error by perhaps 10 to 20 percent or even more. The accuracy of these estimates depends strongly on what reference materials we have available, how much time is allotted for the estimate, and, of course, how experienced we are with similar problems. The first example we present will attempt to illustrate what a professional engineer might be called upon to do in a few minutes with no references. It is not the type of problem you, as a beginning student, would be expected to do because you have not yet accumulated the needed experience.

Example problem 5.1   A civil engineer is asked to meet with a city council committee to discuss their needs with respect to the disposal of solid wastes (garbage or refuse). The community, a city of 12 000 persons, must begin supplying refuse collection and disposal for its citizens for the first time. In reviewing various alternatives for disposal, a sanitary landfill is suggested. One of the council members is concerned about how much land is going to be needed, so she asks the engineer how many acres will be required within the next 10 years.

Discussion   The engineer quickly estimates as follows:
The national average solid waste production is 2.75 kg/capita/day. We can determine that each citizen thus will produce about 1000 kg of refuse per year by the following calculation:

$$(2.75 \text{ kg/day})(365 \text{ days/year}) \cong 1000 \text{ kg/year}$$

Experience indicates that refuse can probably be compacted to a density of 400 to 600 kg/m^3. On this basis the per capita landfill volume will be about 2 m^3 each year; and 1 acre filled 1 m deep will contain the collected refuse of 2000 people for a year (1 acre = 4,047 m^2). Therefore the requirement for 12 000 people will be 1 acre filled 6 m deep. However, knowledge of the geology of the particular area indicates that bedrock occurs at approximately 6 m below the ground surface. The completed landfill, therefore, should have an average depth of 4 m; consequently, 1.5 acres a year, or 15 acres in 10 years, will be required. The patterns of the recent past indicate that some growth in population and solid waste generation should be expected. It is finally suggested that the city should plan to use about 20 acres in the next 10 years.

This calculation took only minutes and required no computational device other than pencil and paper. The engineer's experience, rapid calculations, sound basic assumptions, and sensible rounding of figures were the main requirements. A usable estimate, designed neither to mislead nor to sell a point of view, was provided. If this project proceeds to the actual development of a sanitary landfill, the civil engineer will then gather actual data, refine the calculations, and prepare estimates upon which one would risk a professional reputation.

Example prob. 5.2 is an illustration of a problem that you might be assigned. Here you have the necessary experience to perform the estimation. Not counting the final written presentation, you should be able to do a similar problem in one-half to one hour.

Example problem 5.2   Suppose that your instructor assigns the following problem: Estimate the height of two different flagpoles on your campus. This will be done on a cloudy day so that no shadow is present. The poles are in the ground (not on top of a building) and the bases of the poles are accessible. One of the poles is on level ground and you have available a carpenter's level, straightedge, protractor and masking tape to do the estimate. The other pole is on ground that slopes away from the base and you have available a carpenter's level, straightedge, protractor and a 12 ft tape. See Fig. 5.4 for one student's (whom we will call Dave) response.

Discussion   To estimate the height of the flagpole on level ground, Dave recognizes that he does not have a normal distance measuring device but that he must know some distance in order to use trigonometry to solve the problem. He knows that he is 6′ 1″ (6 ft 1 in) tall and can mark that height on the pole with masking tape, which he can see from a location several feet from the pole. He can use the combined level, protractor and straightedge to estimate angles from the horizontal. The distance away from the pole for

Figure 5.4a

	9 — 24 — XX	ENGR 160 FLAGPOLE PROBLEM	DAVE DOE	1/2

PROBLEM 5.2

ESTIMATE THE HEIGHT OF 2 FLAGPOLES ON YOUR CAMPUS. ASSUME THE SUN IS NOT SHINING AND THAT THE BASES OF THE POLES ARE AT GROUND LEVEL (NOT ON TOP OF A BUILDING) AND THAT THE BASES ARE ACCESSIBLE.

A. DO THE ESTIMATE FOR POLE 1 WHERE YOU HAVE AVAILABLE A CARPENTER S LEVEL, STRAIGHT EDGE, PROTRACTOR AND MASKING TAPE. POLE 1 SITS ON LEVEL GROUND.

B. FOR POLE 2, YOU HAVE A CARPENTER S LEVEL, STRAIGHT EDGE, PROTRACTOR AND A 12' TAPE. THE GROUND AROUND POLE 2 SLOPES AWAY FROM THE BASE.

PART A

ASSUMPTIONS

1. GROUND APPROXIMATELY LEVEL AROUND FLAGPOLE BASE

PROCEDURE

¥ PLACE A PIECE OF MASKING TAPE AT MY HEIGHT ON THE FLAGPOLE.
¥ CHOOSE POSITION ABOUT FLAGPOLE HEIGHT AWAY FROM THE BASE AND MEASURE ELEVATION ANGLES TO TOP OF POLE (β) AND TO TAPE (α). SEE FIG. 1.

DATA

1. MY HEIGHT KNOWN TO BE 6' 1",
2. LEVEL, STRAIGHT EDGE, AND PROTRACTOR CAN SERVE AS A SYSTEM FOR MEASURING ELEVATION ANGLES (FIG. 2)
3. $\alpha = 7.5^\circ$, $\beta = 48^\circ$

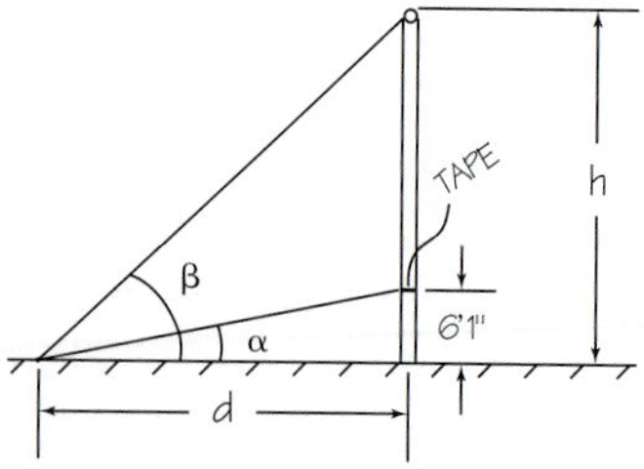

FIGURE 1

SOLUTION

$$\text{TAN } \alpha = \frac{6'1''}{d} = \text{TAN } 7.5¡$$

$$\text{TAN } \beta = \frac{h}{d} = \text{TAN } 48¡$$

$$\therefore h = d(\text{TAN } 48¡) = \left(\frac{6'1''}{\text{TAN } 7.5¡}\right)(\text{TAN } 48¡) = 51.319 \text{ FT}$$

APPROXIMATE $h = 51'$

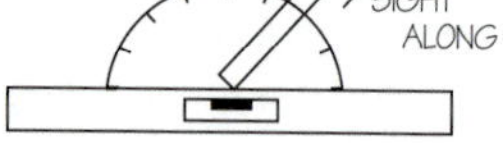

FIGURE 2

Student presentation for Example prob. 5.2.

Figure 5.4b

	9 – 24 – XX	ENGR 160 FLAGPOLE PROBLEM	DAVE DOE	2/2

PART B

ASSUMPTIONS

1. GROUND HAS CONSTANT SLOPE FROM BASE TO MEASURING POINT.

PROCEDURES

- CHOOSE POINT ABOUT FLAGPOLE HEIGHT AWAY FROM BASE AND MEASURE ELEVATION ANGLES TO TOP(γ) AND TO BASE(Δ)
- MEASURE DISTANCE (d) FROM CHOSEN POINT TO BASE. FIG. 3.

DATA

1. $\Delta = 3.5°$
2. $\gamma = 44°$
3. $d = 93'4''$

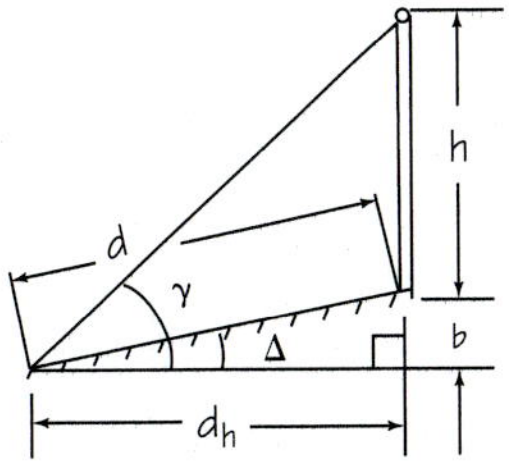

FIGURE 3

SOLUTION

$$\text{SIN } \Delta = \frac{b}{d}$$

$$b = d \text{ SIN } \Delta = (93'4'') \text{ SIN } 3.5° = 5.6979'$$

$$\text{TAN } \Delta = \frac{b}{d_h}$$

$$d_h = \frac{b}{\text{TAN } \Delta} = \frac{5.6979}{\text{TAN } 3.5°} = 93.1598'$$

$$\text{TAN } \gamma = \frac{h+b}{d_h}$$

$$h = d_h \text{ TAN } \gamma - b = (93.1598) \text{ TAN } 44° - 5.6979$$

$$= 84.265'$$

APPROXIMATE $h = 84'$

measuring the angles is arbitrary, but he chooses a distance that will provide angles that are neither too large nor too small, both of which would be difficult to estimate. Then from this point on the ground he estimates the angles to his 6′ 1″ masking tape mark and to the top of the pole. Note that Dave has kept all the significant figures through the calculations and only rounded at the end, reasoning that he does not want intermediate rounding to affect his answer. Based on the method used for estimation, he believes his answer is not closer than the nearest foot so he rounds to that value.

When estimating the height of the pole on sloping ground, Dave had a tape available so it was not necessary to mark his height on the pole. Again, Dave kept all significant figures through the calculation procedure and only rounded the final result.

Example problem 5.3   A homeowner has asked you to estimate the number of gallons of paint required to prime and finish coat her new garage. You are told that paint is applied about 0.004 in thick on smooth surfaces. The siding is to be gray and the roof overhang and trim are to be white. See Fig. 5.5 for one approach done by Laura.

Discussion   From past experience Laura knew that one coat of primer would be needed and that two coats of finish paint would be required for a lasting outcome. She also noted that the garage doors were painted by the manufacturer before installation and therefore would not need further paint. She decided to neglect the effect of windows in the garage because of their small size. She observed that the siding was a rough vertical wood type and that the soffit (underside of the roof overhang) was smooth plywood. Since she had limited experience with rough siding, she contacted a local paint retailer and learned that rough siding would take approximately 3 times as much primer as smooth siding and that the finish paint would cover only about three-fourth of the normal area because of the siding roughness. She carefully documented this fact in her presentation.

Laura took all necessary measurements and computed the areas that must be painted. She determined that the paint film thickness of 0.004 in corresponds to approximately 400 ft^2/gal coverage. Like Dave in the previous example, Laura retained extra significant figures until finally rounding at the end of the estimate. In this case she correctly rounded up rather than to the nearest gallon.

Example problem 5.4   Approximate the amount of gasoline that will be used by the students of Iowa State University during the next break between fall and spring semesters for the purpose of visiting their homes and returning. Provide the answer in gallons. What will be the total cost of this gasoline?

Discussion   Figure 5.6 is the result of the approximation. A number of assumptions were made in order to obtain a solution (without a write-up) in less than 30 min. Note that the write-up was prepared on a word processor.

Figure 5.5a

PROBLEM 5.3

A HOMEOWNER HAS ASKED FOR AN ESTIMATE OF THE NUMBER OF GALLONS OF PAINT REQUIRED TO PRIME AND FINISH COAT HER NEW GARAGE. PAINT SHOULD BE APPLIED ABOUT 0.004 IN. THICK ON SMOOTH SURFACES. THE SIDING IS TO BE GRAY AND THE TRIM WHITE.

ASSUMPTIONS

1. 1 COAT OF PRIMER, 2 FINISH COATS
2. GARAGE DOORS ARE NOT PAINTED.
3. NEGLECT AREA OF SMALL WINDOWS IN GARAGE.

PROCEDURE

MEASURE GARAGE SURFACES TO OBTAIN TOTAL AREA TO BE PAINTED. OBSERVE SMOOTHNESS OF SIDING TO ESTIMATE PAINT COVERAGE. COMPUTE AMOUNT OF EACH TYPE OF PAINT.

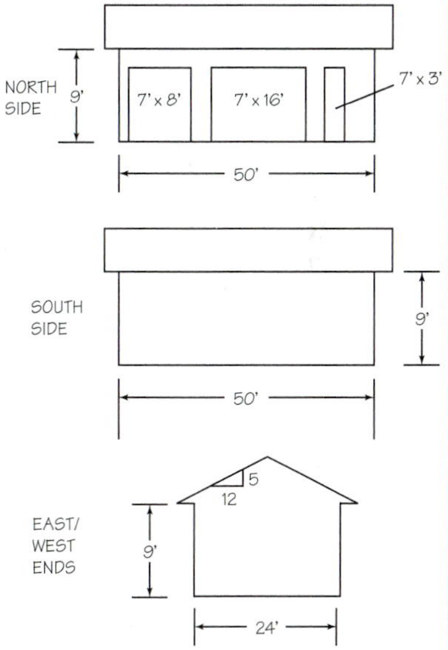

COLLECTED DATA

1. SINCE 1 GAL = 231 IN^3, PAINT THICKNESS OF 0.004 IN RESULTS IN 1 GAL COVERING $\cong$ 400 FT^2 OF SMOOTH SURFACE.
2. OVERHANGS ARE 18 IN.
3. SIDING IS OBSERVED TO BE ROUGH WOOD/VERTICAL. SOFFIT IS SMOOTH PLYWOOD.
4. LOCAL PAINT STORE REPRESENTATIVE SUGGESTED THAT PRIMER ON ROUGH WOOD SIDING COVERS ONLY 1/3 NORMAL AREA AND THAT TOP COAT COVERS ABOUT 3/4 NORMAL AREA.

(Diagram labels: NORTH SIDE, 9', 7' x 8', 7' x 16', 7' x 3', 50'; SOUTH SIDE, 9', 50'; EAST/WEST ENDS, 9', 5, 12, 24')

SOLUTION

NORTH SIDE AREA = $(9)(50) - (7)(8) - (7)(16) - (7)(3) = 261$ FT^2

SOUTH SIDE AREA = $(9)(50) = 450$ FT^2

EAST/WEST END AREA = $9(24) + \frac{1}{2}(24)(5) = 276$ FT^2 /END

OVERHANG AREA = $(53)(1.5)(2) + 2\left[(13.5)^2 + \left\{\left(\frac{5}{12}\right)(13.5)\right\}^2\right]^{1/2}$

$= 159 + 29.25 \cong 188$ FT^2

Student presentation for Example prob. 5.3.

Figure 5.5b

	3 – 16 – XX	ENGR 161 PROBLEM 5.3	LAURA LYNN	2/2

TOTAL AREA OF SIDING = $261 + 450 + 2(276) = 1263$ FT^2

TOTAL OVERHANG AREA = 188 FT^2

PRIMER NEEDED FOR SIDING = $\left(\frac{1263}{400}\right) 3 = 9.47$ GAL

PRIMER NEEDED FOR OVERHANG = $\frac{188}{400} = 0.47$ GAL

TOTAL PRIMER NEEDED = $9.47 + 0.47 = 9.94$ GAL

GRAY FINISH COAT FOR SIDING = $(2)\left(\frac{1263}{400}\right)\left(4/3\right) = 8.42$ GAL

WHITE FINISH COAT FOR OVERHANG/TRIM = $(2)\left(\frac{188}{400}\right) = 0.94$ GAL

RECOMMENDED PURCHASE :

PRIMER: 10 GAL

GRAY TOP COAT: 9 GAL

WHITE TOP COAT: 1 GAL

Figure 5.6a Student presentation, produced on a word processor, for Example prob. 5.4.

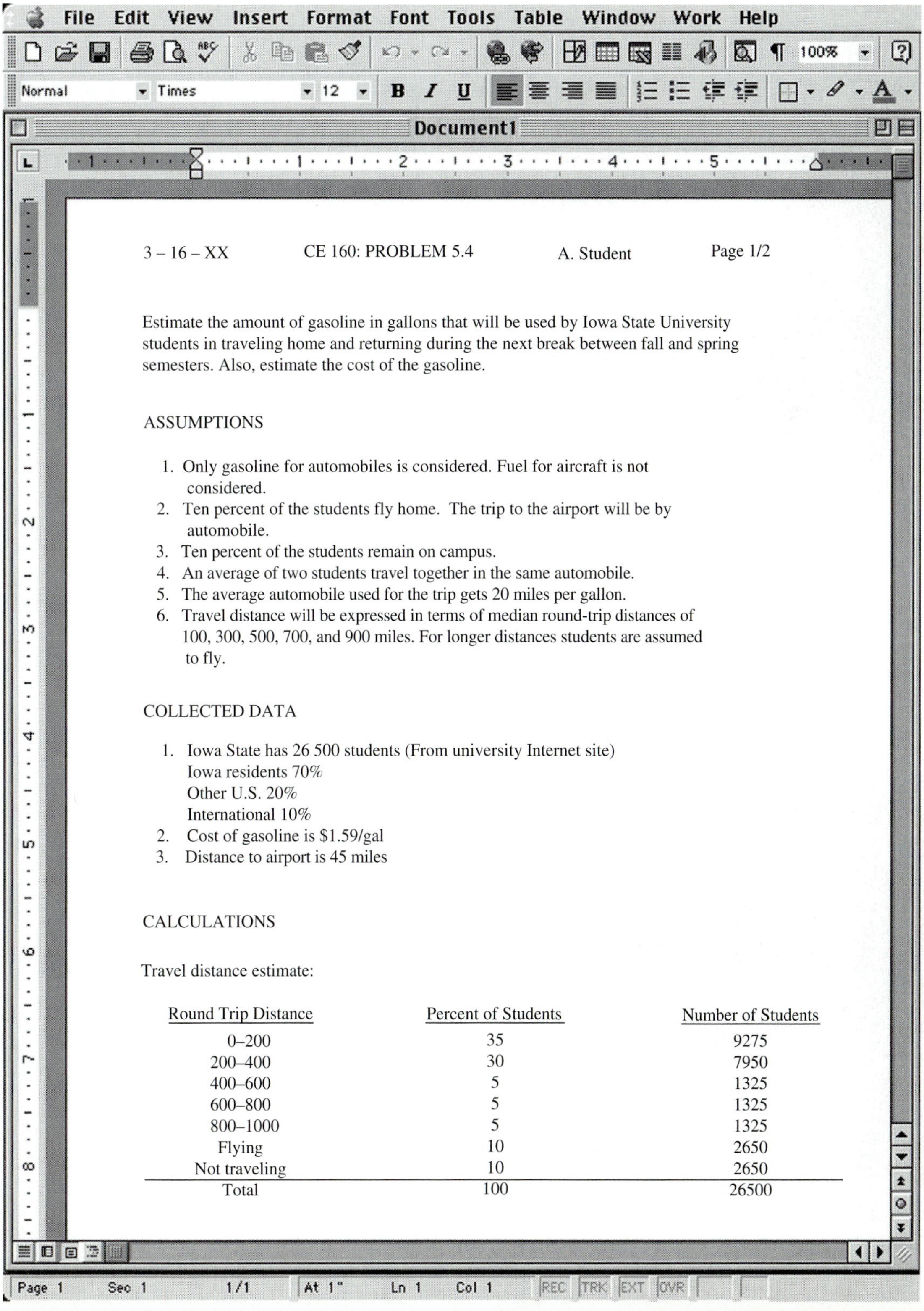

---

3 – 16 – XX   CE 160: PROBLEM 5.4   A. Student   Page 1/2

Estimate the amount of gasoline in gallons that will be used by Iowa State University students in traveling home and returning during the next break between fall and spring semesters. Also, estimate the cost of the gasoline.

ASSUMPTIONS

1. Only gasoline for automobiles is considered. Fuel for aircraft is not considered.
2. Ten percent of the students fly home. The trip to the airport will be by automobile.
3. Ten percent of the students remain on campus.
4. An average of two students travel together in the same automobile.
5. The average automobile used for the trip gets 20 miles per gallon.
6. Travel distance will be expressed in terms of median round-trip distances of 100, 300, 500, 700, and 900 miles. For longer distances students are assumed to fly.

COLLECTED DATA

1. Iowa State has 26 500 students (From university Internet site)
   Iowa residents 70%
   Other U.S. 20%
   International 10%
2. Cost of gasoline is $1.59/gal
3. Distance to airport is 45 miles

CALCULATIONS

Travel distance estimate:

Round Trip Distance	Percent of Students	Number of Students
0–200	35	9275
200–400	30	7950
400–600	5	1325
600–800	5	1325
800–1000	5	1325
Flying	10	2650
Not traveling	10	2650
Total	100	26500

Figure 5.6b

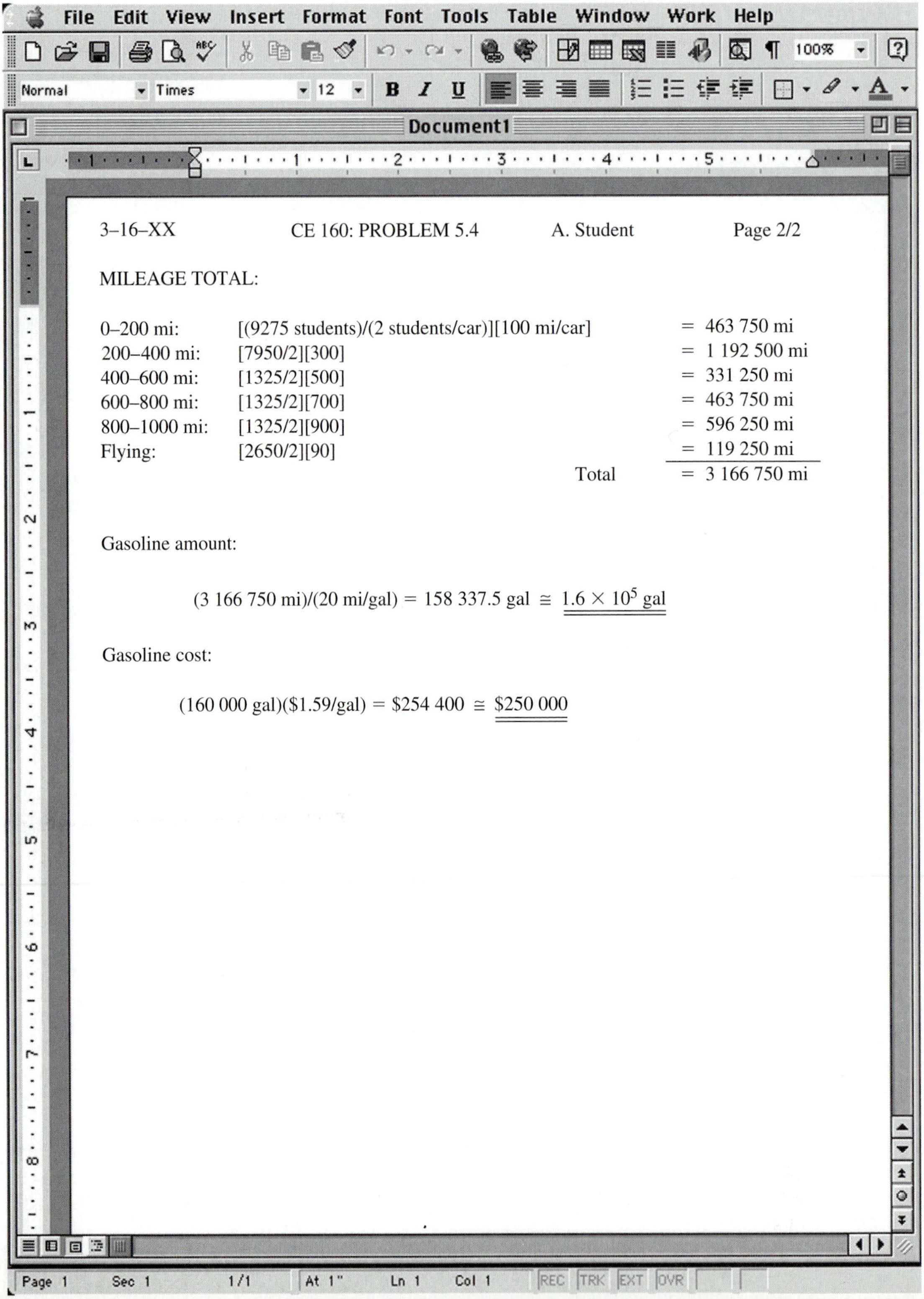

Normal ▾ Times ▾ 12 ▾ **B** *I* U ≣ ≣ ≣ ≣ ⋮≣ ⋮≣ ⋮≣ ⋮≣ ▥ ▾ ✎ ▾ **A** ▾

Document1

L · · · 1 · · · 1 · · · ⧖ · · · 1 · · · 1 · · · 2 · · · 1 · · · 3 · · · 1 · · · 4 · · · 1 · · · 5 · · · 1 · · · △ · · · 1 · ·

3–16–XX   CE 160: PROBLEM 5.4   A. Student   Page 2/2

MILEAGE TOTAL:

0–200 mi:	[(9275 students)/(2 students/car)][100 mi/car]	= 463 750 mi
200–400 mi:	[7950/2][300]	= 1 192 500 mi
400–600 mi:	[1325/2][500]	= 331 250 mi
600–800 mi:	[1325/2][700]	= 463 750 mi
800–1000 mi:	[1325/2][900]	= 596 250 mi
Flying:	[2650/2][90]	= 119 250 mi
	Total	= 3 166 750 mi

Gasoline amount:

$$(3\ 166\ 750\ \text{mi})/(20\ \text{mi/gal}) = 158\ 337.5\ \text{gal} \cong \underline{1.6 \times 10^5\ \text{gal}}$$

Gasoline cost:

$$(160\ 000\ \text{gal})(\$1.59/\text{gal}) = \$254\ 400 \cong \underline{\$250\ 000}$$

The following are some of the terms and concepts you should recognize and understand.

Physical measurement   Random error (in a measurement)
Significant digits   Engineering approximations
Scientific notation   Engineering estimates
Accuracy   Systematic error (in a measurement)
Precision

Problems

5.1 How many significant digits are contained in each of the following quantities?
   (a) 5 535   (f) 5 760 000
   (b) 5 630.0   (g) 222.230
   (c) 0.000 6   (h) 1.320×10^3
   (d) 2 000 000   (i) 2 000 000.0
   (e) 31 700   (j) $4.626\ 7 \times 10^2$

5.2 How many significant digits are contained in each of the following quantities?
   (a) 2.345   (f) 2.54 cm/in
   (b) 0.000 3   (g) 0.000 032
   (c) 0.000 023   (h) 1.000×10^8
   (d) 2 001 000   (i) 60 s/min
   (e) 300 003.0   (j) 4×10^{12}

5.3 Perform the following computations and report with the answer rounded to the proper number of significant digits. No numbers are exact conversions.
   (a) 23.52×372.5 (f) $(1.45 \times 10^4)\,(1.3678 \times 10^{-3})$
   (b) $3.735 - 1.43$ (g) 178.457×753.525
   (c) $6.231\ 827\ (4.23 \times 10^7)$ (h) $17.54678 \div 24.3568$
   (d) $0.25 \div 0.50$ (i) $4\ 300\ 240 \div 784$
   (e) $31 \div 2.0$ (j) $4500.3 + 372$

5.4 Using the values that follow (a) through (f), perform the suggested calculations using exact conversions or with enough significant digits so that it does not affect the accuracy of the answer.
   (a) 4376 feet to miles
   (b) 7.8×10^{10} atoms to moles
   (c) 653.545 kg to N
   (d) 7.358 cm to in
   (e) .625 783 7 · c where c = speed of light
   (f) 253 days to seconds

$$c = 2.997\ 924\ 58 \times 10^8 \text{ m/s} \qquad 1 \text{ mi} = 5\ 280 \text{ ft}$$

$$g = 9.806\ 65 \text{ m/s}^2 \qquad\qquad 1 \text{ day} = 86\ 400 \text{ s}$$

$$N_A = 6.022\ 136\ 736 \times 10^{23} \text{ atoms/mol} \qquad 1 \text{ in} = 2.54 \text{ cm}$$

5.5 Solve the following problems and give the answers rounded to the proper number of significant digits.
   (a) $v = 2.1\,(3.254)t + 2.14t^2$ for $t = 3.2$
   (b) $(24.56)^3$ ft · (12 in/ft) = ? in
   (c) $400 a plate $\times$ 320 guests = $
   (d) $V = [\pi(3.02 \text{ cm})^2\,(7.53 \text{ cm})]/3$ (volume of a cone)
   (e) $325.03 - 527.897 + 615$
   (f) 12¢ per part · 3800 parts

5.6 A pressure gage on an air tank reads 130 lb per square inch (130 lb/in^2). The a note on the face of the gage states ± 3 percent at 130 lb/in^2.

(a) What is the range of air pressure in the tank at 130 lb/in^2?

(b) What is the range of air pressure in the tank at 65 lb/in^2?

5.7 A vacuum gage reads 0.92 kPa. The face of the gage states ± 0.01 kPa.

(a) What is the range of vacuum the gage actually could be?

(b) What is the range when the gage reads 0.98 kPa?

5.8 What is the percent of error if you use a pair of calipers on a 6 in percision gage block and get a reading of 5.998?

5.9 Estimate the cost of taking your classmates to the nearest ice cream shop for their favorite dessert.

5.10 Estimate the number of paper clips that will fit in a box 16 × 10 × 12 in.

5.11 Estimate the number of books you will read in your lifetime.

5.12 Estimate the cost of all the food you will consume this semester.

5.13 Estimate the number of times you will wash a particular pair of socks in the next year.

5.14 Estimate the number of businesses within 1 mile of campus.

5.15 Find the approximate number of pages that are in all the sets of encyclopedias in the library.

5.16 Estimate the cubic feet of natural gas that a specified engineering building will consume for heating next winter.

5.17 Estimate the number of bird's nests that are in the largest tree on campus.

5.18 Get a large bag of raisins. Take out a small sample and use it to estimate the total amount of raisins in the bag. (This also works with M & M's and gummy worms.)

5.19 Estimate the amount of total time spent by the members of this class studying for the class this week.

5.20 If you buy a new pair of shoes at the beginning of the school year and wear them every day until the last day of school, estimate how many miles you will have walked and run in them.

5.21 Estimate the total amount of time you will spend commuting to and from class for this term.

5.22 Estimate the number of gallons of water in the nearest swimming pool.

5.23 Estimate the total of all the times for all the running events for the 2000 summer Olympics.

5.24 Estimate the number of steps on campus.

5.25 Estimate the number of wheelchair ramps on campus and compare it to the answer to Prob. 5.24.

5.26 Estimate the number of garbage bags needed to fill all the trash cans in your campus library.

5.27 Estimate the number of AIDS cases that will be reported next year in Zimbabwe; in all of Africa for the next decade.

5.28 Estimate the amount of money spent on running for government office by congressional candidates in the last election year.

5.29 Estimate the number of homeless in the United States.

5.30 Estimate the number of educators in your state and their average income.

5.31 Estimate the number of people who will be injured but not killed by drunk drivers next year in the United States.

CHAPTER 6

Dimensions, Units, and Conversions

6.1 Introduction

Long ago when countries were more isolated from one another, individual governments tended to develop and use their own set of measures. As the rapid increase in global communication brought countries closer together and the development of technology accelerated, the need for a universal system of measurement has become abundantly clear. There has been an explosion of information but a continued diversity of reporting about these new technologies. It became apparent that a standard set of dimensions, units, and measurements was vital if this wealth of knowledge was to be of benefit to everyone. This chapter deals with the difference between dimensions and units and at the same time explains how there can be an orderly transition from many systems of units to one system—that is, an international standard.

The standard currently accepted in most industrial nations (it is optional in the United States) is the international metric system, or Systeme International d'Unites, abbreviated SI. The SI units are a modification and refinement of an earlier version of the metric system (MKS) that designated the meter, kilogram, and second as fundamental units. (see Fig. 6.1)

France was the first country in 1840, to legislate official adoption of the metric system and decree that its use be mandatory.

The United States almost adopted the metric system 150 years ago. In fact, the metric system was made legal in the United States in 1866, but its use was not compulsory. In spite of many attempts since that time, full conversion to the metric system has not yet been realized in the United States, but significant steps in that direction are continuously underway.

6.2 Physical Quantities

Engineers are constantly concerned with the measurements of fundamental physical quantities such as length, time, temperature, force, and so on. In order to specify a physical quantity fully, it is not sufficient to indicate merely a numerical value. The magnitude of physical quantities can be understood only when they are compared with predetermined reference amounts, called *units*. Any measurement is, in effect, a comparison of how many (a number) units

Figure 6.1

Look for the **km/h** tab below the maximum speed limit sign, indicating that this is the new speed in metric.

Surveillez l'indication de l'unité de vitesse **km/h**; ce symbole signifie que va vitesse est mesurée selon le système métrique.

100 km/h This speed limit will likely be the most common on freeways. On most rural two-lane roadways, **80 km/h** will be typical.

100 km/h Sur les autoroutes, la vitesse maximale la plus courante sera de **100 km/h** tandis que sur les routes à grande circulation, elle sera de **80 km/h**.

50 km/h A **50 km/h** speed limit will apply in most cities. Actual speed limits will be established in accordance with local regulations.

50 km/h Dans la plupart des grands centres, la vitesse maximale sera de **50 km/h**. Les vitesses maximales en vigueur dans votre société seront établies selon les règlements municipaux.

Metric Commission Canada   Commission du système métrique Canada

Commission du système métrique Canada   Metric Commission Canada

Highway signs in Canada. (*Metric Commission of Canada.*)

are contained within a physical quantity. Consider length (L), the physical quantity and 20.0 the numerical value, with meters (m) as the designated units, then a general relation can be represented by the expression

$$\text{Length } (L) = 20.0 \text{ m}$$

For this relationship to be valid the exact reproduction of a unit must be theoretically possible at any time. Therefore standards must be established. These standards are a set of fundamental unit quantities kept under normalized conditions in order to preserve their values as accurately as possible. We will speak more about standards and their importance later.

6.3 Dimensions

Dimensions are used to describe physical quantities; however, the most important consideration to remember is that dimensions are independent of units.

As mentioned in Sec. 6.1, the physical quantity length can be represented by the dimension L, for which there are a large number of possibilities available when selecting a unit. For example, in ancient Egypt the cubit was a unit related to the length of the arm from the tip of the middle finger to the elbow. Measurements thus were a function of physical stature, with variation from one individual to another. Much later, in Britain, the inch was specified as the distance covered by three barley corns, round and dry, laid end to end.

Today we require more precision. For example, the meter is defined in terms of the distance traveled by light in a vacuum during a specified amount of time. We can draw two important points from this discussion: (1) Physical quantities can be accurately measured and (2) each of these units (cubit, inch, and meter), although distinctly different, has in common the quality of being a length and not an area or a volume.

A technique used to distinguish between units and dimensions is to call all physical quantities of length a dimension (e.g., L). In this way each new physical quantity gives rise to a new dimension, such as T for time, F for force, M for mass, and so on. (*Note*: There are as many dimensions as there are kinds of physical quantities.)

However, to simplify the process, dimensions are divided into two areas: fundamental and derived. A fundamental dimension is a dimension that can be conveniently and usefully manipulated when expressing all physical quantities of a particular field of science or engineering. Derived dimensions are a combination of fundamental dimensions. Velocity, for example, could be defined as fundamental dimension V, but it is more customary as well as more convenient to consider velocity as a combination of fundamental dimensions so that it becomes a derived dimension, $V = (L)(T)^{-1}$. L and T are fundamental dimensions, and V is a derived dimension because it is made up of two fundamental dimensions (L, T).

It is advantageous to use as few fundamental dimensions as possible, but the selection of what is to be fundamental and what is to be derived is not fixed. In actuality, any dimension can be selected as a fundamental dimension in a particular field of engineering or science; and for reasons of convenience it may be a derived dimension in another field.

Once a set of primary dimensions have been adopted, a base unit for each primary dimension must then be specified.

A *dimensional system* can be defined as the smallest number of fundamental dimensions which will form a consistent and complete set for a field of science. For example, three fundamental dimensions are necessary to form a complete mechanical dimensional system. Depending on the discipline, these dimensions may be specified as either length (L), time (T), and mass (M) or length (L), time (T), and force (F). If temperature is important to the application, a fourth fundamental dimension may be added.

The *absolute system* (so called because dimensions used are not affected by gravity) has as its fundamental dimensions L, T, and M. An advantage of this system is that comparisons of masses at various locations can be made with an ordinary balance because the local acceleration of gravity has no influence upon the results.

The *gravitational system* has as its fundamental dimensions L, T, and F. It is widely used in many engineering branches because it simplifies computations when weight is a fundamental quantity in the computations. Table 6.1

Table 6.1 Two Basic Dimensional Systems

Quantity	Absolute	Gravitational
Length	L	L
Time	T	T
Mass	M	$FL^{-1}T^2$
Force	MLT^{-2}	F
Velocity	LT^{-1}	LT^{-1}
Pressure	$ML^{-1}T^{-2}$	FL^{-2}
Momentum	MLT^{-1}	FT
Energy	ML^2T^{-2}	FL
Power	ML^2T^{-3}	FLT^{-1}
Torque	ML^2T^{-2}	FL

illustrates two of the more basic dimensional systems; however, a number of other dimensional systems are commonly used depending on the specific discipline.

6.4 Units

After a consistent dimensional system has been selected, the next step would be to select a specific unit for each fundamental dimension. The problem one encounters when working with units is that there can be a large number of unit systems from which to choose for any given dimensional system. It is obviously desirable to limit the number of systems and combinations of systems. The SI previously alluded to is intended to serve as an international standard that will provide worldwide consistency.

Two fundamental systems of units are commonly used in mechanics today. One system used in almost every industrial country of the world is called the *metric system.* It is a decimal, absolute system based on the meter, kilogram, and second (MKS) as the units of length, mass, and time, respectively. The United States has traditionally used the Engineering System, which is based on the foot, pound-force, and second.

Numerous international conferences on weights and measures over the past 40 years have gradually modified the MKS system to the point that all countries previously using various forms of the metric system are beginning to standardize. The Système International d'Unités (SI) is now considered the new international system of units. The United States has offically adopted this system as indicated earlier, but full implementation will be preceded by a long and expensive period of change. During this transition period engineers will have not only to be familiar with SI but also to know other systems and the necessary conversion process between or among systems. This chapter will devote a high percentage of content to the international standard (SI units and symbols); however, some examples and explanations of the Engineering System and the U.S. Customary System (also called the British Gravitational System) will be included.

The International System of Units (SI), developed and maintained by the General Conference on Weights and Measures (Conference Generale des Poids et Mesures, CGPM), is intended as a basis for worldwide standardization of measurements. The name and abbreviation were set forth in 1960. SI at the present time is a complete system that is being universally adopted.

This new international system is divided into three classes of units:

1. Base units
2. Supplementary units
3. Derived units

There are seven base units in the SI. The units (except the kilogram) are defined in such a way so that they can be reproduced anywhere in the world.

Table 6.2 lists each base unit along with its name and proper symbol.

In the following list each of the base units is defined as established at the international CGPM:

1. *Length.* The meter (m) is a length equal to the distance traveled by light in a vacuum during $1/299\ 792\ 458$ s. The meter was defined by the CGPM that met in 1983.
2. *Time.* The second (s) is the duration of $9\ 192\ 631\ 770$ periods of radiation corresponding to the transition between the two hyperfine levels of the ground state of the cesium-133 atom. The second was adopted by the thirteenth CGPM in 1967.
3. *Mass.* The standard for the unit of mass, the kilogram (kg), is a cylinder of platinum-iridium alloy kept by the International Bureau of Weights and Measures in France. A duplicate copy is maintained in the United States. The unit of mass was adopted by the First and Third CGPMs in 1889 and 1901. It is the only base unit nonreproducible in a properly equipped laboratory.
4. *Electric current.* The ampere (A) is a constant current which, if maintained in two straight parallel conductors of infinite length and of negligible circular cross sections and placed one meter apart in volume, would produce between these conductors a force equal to 2×10^{-7} newton per meter of length. The ampere was adopted by the Ninth CGPM in 1948.
5. *Temperature.* The kelvin (K), a unit of thermodynamic temperature, is the fraction $1/273.16$ of the thermodynamic temperature of the triple point of water. The kelvin was adopted by the Thirteenth CGPM in 1967.

Table 6.2 Base Units

Quantity	Name	Symbol
Length	meter	m
Mass	Kilogram	kg
Time	second	s
Electric current	ampere	A
Thermodynamic temp.	kelvin	K
Amount of substance	mole	mol
Luminous intensity	candela	cd

Table 6.3 Supplementary Units

Quantity	Name	Symbol
Plane angle	radian	rad
Solid angle	steradian	sr

6. *Amount of substance.* The mole (mol) is the amount of substance of a system that contains as many elementary entities as there are atoms in 0.012 kilogram (kg) of carbon-12. The mole was defined by the Fourteenth CGPM in 1971.

7. *Luminous intensity.* The base unit candela (cd) is the luminous intensity in a given direction of a source that emits monochromatic radiation of frequency 540×10^{12} hertz (Hz) and has a radiant intensity in that direction of $1/683$ watts per steradian (W/sr).

The units listed in Table 6.3 are called *supplementary units* and may be regarded as either base units or derived units.

The unit for a plane angle is the radian, a unit that is used frequently in engineering. The steradian is not as commonly used. These units can be defined in the following way:

1. *Plane angle.* The radian (rad) is the plane angle between two radii of a circle that cuts off on the circumference of an arc equal in length to the radius.
2. *Solid angle.* The steradian (sr) is the solid angle which, having its vertex in the center of a sphere, cuts off an area of the sphere equal to that of a square with sides of length equal to the radius of the sphere.

As indicated earlier, derived units are formed by combining base, supplementary, or other derived units. Symbols for them are carefully selected to avoid confusion. Those which have special names and symbols, as interpreted for the United States by the National Bureau of Standards, are listed in Table 6.4 together with their definitions in terms of base units.

At first glance Fig. 6.2 may appear complex, even confusing; however, a considerable amount of information is presented in this concise flowchart. To get the point quickly, be aware that the solid lines denote multiplication and the broken lines indicate division. The arrows pointing toward the units (circled) are significant and arrows going away have no meaning for that particular unit. Consider the pascal, as an example: Two arrows point toward the circle—one solid and one broken. This means that the unit pascal is formed from the newton and meter squared, or N/m^2.

Other derived units, such as those included in Table 6.5, have no special unit names but are combinations of base units and units with special names.

Being a decimal system, the SI is convenient to use because by simply affixing a prefix to the base, a quantity can be increased or decreased by factors of 10 and the numerical quantity can be kept within manageable limits. Table 6.6 lists the multiplication factors with their prefix names and symbols.

The proper selection of prefixes also will help to eliminate nonsignificant zeros and leading zeros in decimal fractions. One rule to follow is that the numerical value of any measurement should be recorded as a number between

Table 6.4 Derived Units

Quantity	SI Unit Symbol	Name	Base Units
Frequency	Hz	hertz	s^{-1}
Force	N	newton	$kg \cdot m \cdot s^{-2}$
Pressure stress	Pa	pascal	$kg \cdot m^{-1} \cdot s^{-2}$
Energy or work	J	joule	$kg \cdot m^2 \cdot s^{-2}$
Quantity of heat	J	joule	$kg \cdot m^2 \cdot s^{-2}$
Power radiant flux	W	watt	$kg \cdot m^2 \cdot s^{-3}$
Electric charge	C	coulomb	$A \cdot s$
Electric potential	V	volt	$kg \cdot m^2 \cdot s^{-3} \cdot A^{-1}$
Potential difference	V	volt	$kg \cdot m^2 \cdot s^{-3} \cdot A^{-1}$
Electromotive force	V	volt	$kg \cdot m^2 \cdot s^{-3} \cdot A^{-1}$
Capacitance	F	farad	$A^2 \cdot s^4 \cdot kg^{-1} \cdot m^{-2}$
Electric resistance	Ω	ohm	$kg \cdot m^2 \cdot s^{-3} \cdot A^{-2}$
Conductance	S	siemens	$kg^{-1} \cdot m^{-2} \cdot s^3 \cdot A^2$
Magnetic flux	Wb	weber	$kg^{-1} \cdot m \cdot s^{-2} \cdot A^{-1}$
Magnetic flux density	T	tesla	$kg \cdot s^2 \cdot A^{-1}$
Inductance	H	henry	$kg \cdot m^2 \cdot s^{-2} \cdot A^{-2}$
Luminous flux	lm	lumen	$cd \cdot sn$
Illuminance	lx	lux	$cd \cdot sn \cdot m^{-2}$
Celsius temperature*	C	degree Celsius	K
Activity (radionuclides)	Bq	becqueret	s^{-1}
Absorbed dose	Gy	gray	$m^2 \cdot s^{-2}$
Dose equivalent	S	sievert	$m^2 \cdot s^{-2}$

*The thermodynamic temperature (T_K) expressed in kelvins is related to Celsius temperature (t_C) expressed in degrees Celsius by the equation $t_C = T_K - 273.15$.

Table 6.5 Common Derived Units

Quantity	Units	Quantity	Units
Acceleration	$m \cdot s^{-2}$	Molar entropy	$J \cdot mol^{-1} \cdot K^{-1}$
Angular acceleration	$rad \cdot s^{-2}$	Molar heat capacity	$J \cdot mol^{-1} \cdot K^{-1}$
Angular velocity	$rad \cdot s^{-1}$	Moment of force	$N \cdot m$
Area	m^2	Permeability	$H \cdot m^{-1}$
Concentration	$mol \cdot m^{-3}$	Permittivity	$F \cdot m^{-1}$
Current density	$A \cdot m^{-2}$	Radiance	$W \cdot m^{-2} \cdot sr^{-1}$
Density, mass	$kg \cdot m^{-3}$	Radiant intensity	$W \cdot sr^{-1}$
Electric charge density	$C \cdot m^{-3}$	Specific heat capacity	$J \cdot kg^{-1} \cdot K^{-1}$
Electric field strength	$V \cdot m^{-1}$	Specific energy	$J \cdot kg^{-1}$
Electric flux density	$C \cdot m^{-2}$	Specific entropy	$J \cdot kg^{-1} \cdot K^{-1}$
Energy density	$J \cdot m^{-3}$	Specific volume	$m^3 \cdot kg^{-1}$
Entropy	$J \cdot K^{-1}$	Surface tension	$N \cdot m^{-1}$
Heat capacity	$J \cdot K^{-1}$	Thermal conductivity	$W \cdot m^{-1} \cdot K^{-1}$
Heat flux density	$W \cdot m^{-2}$	Velocity	$m \cdot s^{-1}$
Irradiance	$W \cdot m^{-2}$	Viscosity, dynamic	$Pa \cdot s$
Luminance	$cd \cdot m^{-2}$	Viscosity, kinematic	$m^2 \cdot s^{-1}$
Magnetic field strength	$A \cdot m^{-1}$	Volume	m^3
Molar energy	$J \cdot mol^{-1}$	Wavelength	m

Figure 6.2

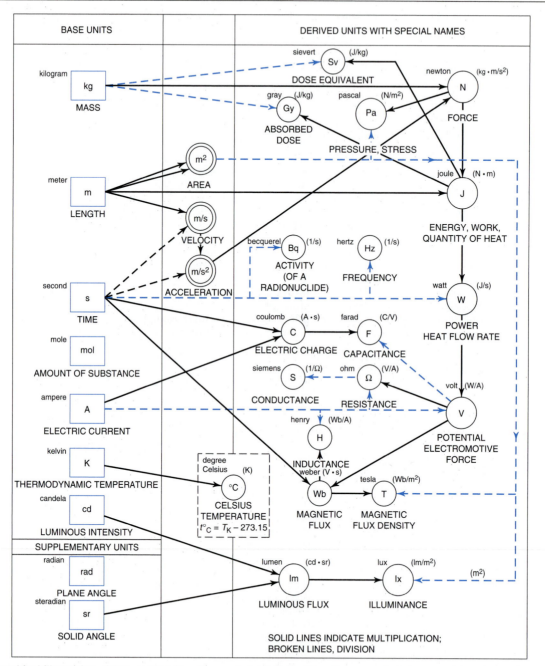

Graphical illustration of how certain SI units are derived in a coherent manner from base and supplementary units. (*National Bureau of Standards.*)

Table 6.6 Decimal Multiples

Multiplier	Prefix name	Symbol
10^{18}	exa	E
10^{15}	peta	P
10^{12}	tera	T
10^{9}	giga	G
10^{6}	*mega	M
10^{3}	*kilo	k
10^{2}	hecto	h
10^{1}	deka	da
10^{-1}	deci	d
10^{-2}	centi	c
10^{-3}	*milli	m
10^{-6}	*micro	μ
10^{-9}	nano	n
10^{-12}	pico	p
10^{-15}	femto	f
10^{-18}	atto	a

*Most often used.

0.1 and 1 000. This rule is suggested because it is easier to make realistic judgments when working with numbers between 0.1 and 1 000. For example, suppose that you are asked the distance to a nearby town. It would be more understandable to respond in kilometers than in meters. That is, it is easier to visualize 10 km than 10 000 m.

Moreover, the use of certain prefixes is preferred over that of others. Those representing powers of 1 000, such as kilo-, mega-, milli-, and micro-, will reduce the number you must remember. These preferred prefixes should be used, with the following three exceptions, which are still common because of convention:

1. When expressing area and volume, the prefixes hecto-, deka-, deci-, and centi- may be used, for example, cubic centimeter.
2. When discussing different values of the same quantity or expressing them in a table, calculations are simpler to perform when you use the same unit multiple throughout.
3. Sometimes a particular multiple is recommended as a consistent unit although its use violates the 0.1 to 1 000 rule. For example, many companies use the millimeter for linear dimensions even when the values lie far outside this suggested range. The cubic decimeter (commonly called liter) also is used.

Recalling the discussion of significant figures, we see that SI prefix notations can be used to a definite advantage.

Consider the previous example of 10 km. When giving an estimate of distance to the nearest town, there is certainly an implied approximation in the use of a round number. Suppose that we were talking about a 10 000 m Olympic track and field event. The accuracy of such a distance must certainly be greater than something between 5 000 and 15 000 m. This example is intended to illustrate the significance of the four zeros. If all four zeros are, in fact, significant, then the race is accurate within 1 m (9 999.5 to 10 000.5). If only three zeros are significant, then the race is accurate to within 10 m (9 995 to 10 005).

There are two logical and acceptable methods available that eliminate confusion concerning zeros or the correct number of significant figures:

1. Use proper prefixes to denote intended significance.

Distance	Precision
10.000 km	9 999.5 to 10 000.5 m
10.00 km	9 995 to 10 005 m
10.0 km	9 950 to 10 050 m
10 km	5 000 to 15 000 m

2. Use scientific notation to indicate the significance.

Distance	Precision
10.000×10^3 m	9 999.5 to 10 000.5 m
10.00×10^3 m	9 995 to 10 005 m
10.0×10^3 m	9 950 to 10 050 m
10×10^3 m	5 000 to 15 000 m

Selection of a proper prefix is customarily the logical way to handle problems of significant figures; however, there are conventions that do not lend themselves to the prefix notation. An example would be temperature in degrees Celsius; that is, $4.00(10^3)$C is the conventional way to handle it, not 4.00 kC.

6.6 Rules for Using SI Units

Along with the adoption of SI comes the responsibility to understand thoroughly and to apply properly the new system. Obsolete practices involving both English and metric units are widespread. This section provides rules that should be followed when working with SI units.

6.6.1 Unit Symbols and Names

1. Periods are never used after symbols unless the symbol is at the end of a sentence (i.e., SI unit symbols are not abbreviations).
2. Unit symbols are written in lowercase letters unless the symbol derives from a proper name, in which case the first letter is capitalized.

Lowercase	Uppercase
m, kg, s, mol, cd	A, K, Hz, Pa, C

3. Symbols rather than self-styled abbreviations always should be used to represent units.

Correct	Not Correct
A	amp
s	sec

4. An s is never added to the symbol to denote plural.
5. A space is always left between the numerical value and the unit symbol.

Correct	Not Correct
43.7 km	43.7km
0.25 Pa	0.25Pa

Exception: No space should be left between numerical values and the symbols for degree, minute, and second of angles and for degree Celsius.

6. There should be no space between the prefix and the unit symbols.

Correct	Not Correct
mm, MΩ	k m, μ F

7. When writing unit names, lowercase all letters except at the beginning of a sentence, even if the unit is derived from a proper name.
8. Plurals are used as required when writing unit names. For example, henries is plural for henry. The following exceptions are noted:

Singular	Plural
lux	lux
hertz	hertz
siemens	siemens

With these exceptions, unit names form their plurals in the usual manner.

9. No hyphen or space should be left between a prefix and the unit name. In three cases the final vowel in the prefix is omitted: megohm, kilohm, and hectare.
10. The symbol should be used in preference to the unit name because unit symbols are standardized. An exception to this is made when a number is written in words preceding the unit, for example, we would write *ten meters*, not *ten m*. The same is true the other way, for example, 10 m, not 10 meters.

6.6.2 Multiplication and Division

1. When writing unit names as a product, always use a space (preferred) or a hyphen.

   Correct Usage
   newton meter or newton-meter

2. When expressing a quotient using unit names, always use the word *per* and not a solidus (/). The solidus, or slash mark, is reserved for use with symbols.

Correct	Not Correct
meter per second	meter/second

3. When writing a unit name that requires a power, use a modifier, such as squared or cubed, after the unit name. For area or volume the modifier can be placed before the unit name.

Correct	Not Correct
millimeter squared	square millimeter

4. When expressing products using unit symbols, the center dot is preferred.

   Correct
   N · m for newton meter

5. When denoting a quotient by unit symbols, any of the following methods are accepted form:

   Correct

   $$\text{m/s or m} \cdot \text{s}^{-1} \text{ or } \frac{\text{m}}{\text{s}}$$

In more complicated cases negative powers or parentheses should be considered. Use m/s^2 or $\text{m} \cdot \text{s}^{-2}$, but not m/s/s for acceleration; use $\text{kg} \cdot \text{m}^2/(\text{s}^3 \cdot \text{A})$ or $\text{kg} \cdot \text{m}^2 \cdot \text{s}^{-3} \cdot \text{A}^{-1}$, but not $\text{kg} \cdot \text{m}^2/\text{s}^3/\text{A}$ for electric potential.

6.6.3 Numbers

1. To denote a decimal point, use a period on the line. When expressing numbers less than 1, a zero should be written before the decimal.

   Example
   15.6
   0.93

2. Since a comma is used in many countries to denote a decimal point, its use is to be avoided in grouping data. When it is desired to avoid this confusion, recommended practice calls for separating the digits into groups of three, counting from the decimal to the left or right, and using a small space to separate the groups.

   Correct and Recommended Procedure
   6.513 824   76 851   7 434   0.187 62

6.6.4 Calculating with SI Units

Before looking at some suggested procedures that will simplify calculations in SI, we should review the following positive characteristics of the system.

Only one unit is used to represent each physical quantity, such as the meter for length, the second for time, and so on. The SI metric units are *coherent;* that is, each new derived unit is a product or quotient of the fundamental and supplementary units without any numerical factors. Since coherency is a strength of the SI system, it would be worthwhile to demonstrate this characteristic by using two examples. First, consider the use of the newton as the unit of force. It is defined by Newton's second law, $F \propto ma$, and it is important to understand that the newton was *defined* as the magnitude of a force needed to impart an acceleration of one meter per second squared 1.0 m/s^2 to a mass of one kilogram (1.0 kg). Thus

$$1.0 \text{ N} = (1.0 \text{ kg})(1.0 \text{ m/s}^2)$$

Table 6.7 Non-SI Units Accepted for Use in the United States

Quantity	Name	Symbol	SI equivalent
Time	minute	min	60 s
	hour	h	3 600 s
	day	d	86 400 s
Plane angle	degree	°	$\pi/180$ rad
	minute	′	$\pi/10\ 800$ rad
	second	″	$\pi/648\ 000$ rad
Volume	liter	L*	10^{-3} m^3
Mass	metric ton	t	10^3 kg
	unified atomic mass unit	u	$1.660\ 57 \times 10^{-27}$ kg (approx)
Land area	hectare	ha	10^4 m^2
Energy	electronvolt	eV	1.602×10^{-19} J (approx)

*Both "L" and "l" are acceptable international symbols for liter. The uppercase letter is recommended for use in the United States because the lowercase "l" can be confused with the numeral 1.

Applying Newton's second law, we can write the equation as equality by the following method:

$$F = \frac{ma}{g_C}, \text{ where } g_C = \frac{ma}{F} \quad \text{or} \quad g_C = \frac{1.0 \text{ kg} \cdot 1.0 \text{ m}}{N \cdot s^2}$$

This constant of proportionality (g_C) serves as a reminder that the units are, in fact, consistent and that the conversion factor is 1.0.

Consider the Joule next, the SI equivalent for the British thermal unit (Btu), calorie, foot-pound-force, electronvolt, and horsepower-hour intended to represent any form of energy. It is defined as the amount of work done when an applied force of one newton (1.0 N) acts through a distance of one meter (1.0 m) in the direction of the force. Thus

$$1.0 \text{ J} = (1.0 \text{ N})(1.0 \text{ m})$$

To maintain the coherency of units, however, we must express time in seconds rather than in minutes or in hours, since the second is the base unit. Once coherency is violated, then a conversion factor must be included and the advantage of the system is diminished.

But there are certain units *outside* SI that are accepted for use in the United States, even though they diminish the system's coherence. These exceptions are listed in Table 6.7.

Calculations using SI can be simplified if you

1. Remember that fundamental relationships are simple and easier to use because of coherence.
2. Recognize how to manipulate units and to gain a proficiency in doing so. Since Watt = J/s = N · m/s, you should realize that N · m/s = (N/m^2)(m^3/s) = (pressure)(volume flow rate).
3. Understand the advantage of occasionally adjusting all variables to base units, replacing N with kg · m/s^2, Pa with kg · m^{-1} · s^{-2}, and so on.

4. Develop a proficiency with exponential notation of numbers to be used in conjunction with unit prefixes.

$$1 \text{ mm}^3 = (10^{-3} \text{ m})^3 = 10^{-9} \text{ m}^3$$

$$1 \text{ ns}^{-1} = (10^{-9} \text{ s})^{-1} = 10^9 \text{ s}^{-1}$$

When calculating with SI, the term "weight" can be confusing. Frequently we hear statements such as "The person weighs 100 kg." A correct statement would be "The person has a mass of 100 kg." To clarify any confusion, let us look at some basic definitions.

First, the term "mass" should be used to indicate only a quantity of matter. Mass, as we know, is measured in kilograms (kg) or pound-mass (lbm) and is always measured against a standard.

Force, as defined by the International Standard of Units (SI), is measured in newtons. By definition it was established as the force required to accelerate a mass of one kilogram to one meter per second squared.

The acceleration of gravity varies at different points on the surface of the earth as well as distance from the earth's surface. The accepted standard value of gravitational acceleration is 9.806 650 m/s^2 at sea level and 45° (degrees) latitude.

Gravity is instrumental in measuring mass with a balance or scale. If you use a beam balance to compare an unknown quantity against a standard mass, the effect of gravity on the two masses cancels out. If you use a spring scale, mass is measured indirectly since the instrument responds to the local force of gravity. Such a scale can be calibrated in mass units and can be reasonably accurate when used where the variation in the acceleration of gravity is not significant.

The following example problem clarifies the confusion that exists in the use of the term "weight" to mean either force or mass. In everyday use the term "weight" nearly always means mass; thus when a person's weight is discussed, the quantity referred to is mass.

Example problem 6.1 A "weight" of 100.0 kg (the unit itself indicates mass) is suspended by a rope (see Fig. 6.3). Calculate the tension in the rope in newtons to hold the mass stationary when the local gravitational acceleration is (*a*) 9.807 m/s^2 and (*b*) 1.63 m/s^2 (approximate value for the surface of the moon).

Theory Tension in the rope or force required to hold the object when the mass is at rest or moving at constant velocity is

$$F = \frac{mg_L}{g_C}$$

where g_L replaces a and is the local acceleration of gravity, g_C is the proportionality constant, and m is the mass of the object. Remember, due to coherence

$$g_C = \frac{1.0 \text{ kg} \cdot \text{m}}{\text{N} \cdot \text{s}^2}$$

Figure 6.3

215

U.S. Customary and Engineering Systems

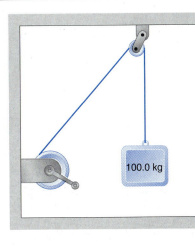

Assumption   Neglect the mass of the rope.

Solution

(*a*) For $g_L = 9.807 \text{ m/s}^2$ (given to four significant figures)

$$F = (100.0 \text{ kg})(9.807 \text{ m/s}^2)/1.0 \text{ kg} \cdot \text{m/N} \cdot \text{s}^2 = 980.7 \text{ N}$$

$$= 0.980 \ 7 \text{ kN}$$

(*b*) For $g_L = 1.63 \text{ m/s}^2$

$$F = (100.0 \text{ kg})(1.63 \text{ m/s}^2)/1.0 \text{ kg} \cdot \text{m/N} \cdot \text{s}^2$$

$$= 0.163 \text{ kN}$$

6.7   U.S. Customary and Engineering Systems

Although the SI ultimately is intended to be adopted worldwide, at the present time many segments of the U.S. industrial complex regularly use other systems. For many years to come engineers in the United States will have to be comfortable and proficient with a variety of other systems.

    There are two systems of units other than SI that are commonly used in the United States. The first, the U.S. Customary System (formerly known as the British Gravitational System), has the fundamental units, foot (ft) for length, pound (lb) for force, and second (s) for time (see Table 6.8). In this system mass (m) is a new derived unit and is called the slug. As a new derived unit, its magnitude can be established. A *slug* is defined as a specific amount of mass: It is the amount of mass that would be accelerated to one foot per second squared given a force of one pound. This system works perfectly well as long as mass is derived totally independent of force. In fact, since we define the derived unit

Table 6.8   The U.S. Customary System

Quality	Unit	Symbol
Mass	slug	slug
Length	foot	ft
Time	second	s
Force	pound	lb

mass as the slug and establish its mass as a quantity of matter that will be accelerated to 1.0 ft/s² when a force of 1.0 lb is applied, we have a coherent system of units. We once again can write Newton's second law as an equality.

$$F = \frac{ma}{g_C}, \text{ where } g_C = \frac{ma}{F} \quad \text{or} \quad g_C = \frac{1.0 \text{ slug} \cdot 1.0 \text{ ft}}{\text{lbf} \cdot \text{s}^2}$$

The second system, called the Engineering System, will be considered next (see Table 6.9). Once again, length, time, and force are selected as the fundamental dimensions. In the Engineering System the pound is used to denote both mass and force. To avoid confusion a convention has been established that assigns (lbm) for pound-mass and (lbf) for pound-force.

In the Engineering System force is specified as a fundamental dimension and its unit (lbf) is adopted independent of its other base units length (ft) and time (s). However, one problem of eternal confusion is about to be presented.

Sometime during the fourtheen century the unit of mass (lbm) was established. At that time it also was an established convention to consider the pound-force (lbf) as the effort required to hold a one pound-mass elevated in a gravitational field where the local acceleration of gravity was 32.1740 ft/s². Since the magnitude of the unit associated with the fundamental dimension (force) had been specified and generally accepted for many years the unit associated with the force dimension was not changed. No problem so far.

However, when the derived unit of mass also was left as the long-standing, established tradition, we must prepare ourselves for some confusion. The derived unit of mass in the Engineering System is not the slug (which is 32.1740 times as large as the pound-mass) but is the traditional pound-mass (lbm). This historical event requires that we include a conversion factor called the constant of proportionality (g_C). For this reason we included (g_C) in most equations that relate pound-mass and pound-force. It provides us with a visual reminder that

$$g_C = 32.1740 \ \frac{\text{lbm} \cdot \text{ft}}{\text{lbf} \cdot \text{s}^2}$$

Table 6.9   The Engineering System

Quality	Unit	Symbol
Mass	pound-mass	lbm
Length	foot	ft
Time	second	s
Force	pound-force	lbf

Since the pound-mass was not independently specified as a derived unit but was retained as a historical amount, Sir Issac Newton's second law of mechanics has to be written as proportional, that is, $F \propto ma$. Newton (1642–1727) did not develop his laws of mechanics until 200 to 300 years after mass and force were defined.

It turns out that a force of 1.0 lbf is quite large. In fact, it is large enough to provide an acceleration of 1.0 ft/s^2 to a mass of 32,174 0 lbm, or an acceleration of 32,174 0 ft/s^2 to a mass of 1.0 lbm.

Once again, when you are calculating in the Engineering System, the constant of proportionality g_C and the local gravitational constant g_L can be particularly confusing when using the term "weight." If you were to hold a small child in your arms, you might say that this child is heavy, and ask the question, how much does this child weigh? Does the question refer to the amount of force exerted to hold the child (lbf) or the child's mass (lbm)?

Normally the term "weight" refers to pound-mass. In other words, the child's mass is 50.0 lbm. The child only requires a force of 50.0 lbf to hold where the local acceleration of gravity is exactly 32,174 0 ft/s^2.

$$F = \frac{mg_L}{g_C} = \frac{50.0 \text{ lbm}}{1.0} \frac{32.174 \text{ ft}}{s^2} \frac{\text{lbf} \cdot s^2}{32{,}174 \text{ 0 lbm} \cdot \text{ft}} = 50.0 \text{ lbf}$$

If the local gravitational constant were any value other than 32,174 0, then the force required to hold the child would be either greater than or less than the force required in the example, for instance, on planet X the local acceleration of gravity is 8.72 ft/s^2. In this case the force required to hold the 50.0 lbm child (mass never changes) would be determined as follows:

$$F = \frac{mg_L}{g_C} = \frac{50.0 \text{ lbm}}{1.0} \frac{8.72 \text{ ft}}{s^2} \frac{\text{lbf} \cdot s^2}{32{,}174 \text{ 0 lbm} \cdot \text{ft}} = 13.6 \text{ lbf}$$

So the next time someone asks how much you can bench press, say about 600 lbm, just don't mention on what planet.

Once again, a word of caution when using the Engineering System in derived expressions such as Newton's second law ($F \propto ma$). This particular combination of units, lbf, lbm, and ft/s^2, are not a coherent set. That is, 1.0 lbf imparts an acceleration of 32.174 ft/s^2 to 1.0 lbm rather than 1.0 ft/s^2 required for coherency. A coherent set of non-SI units involves lbf, slug, and ft/s^2 because 1.0 lbf is required to accelerate 1.0 slug to 1.0 ft/s^2. You always can convert mass quantities from lbm to slugs before substituting into $F = \dfrac{ma}{g_C}$.

The slug is 32.174 times the size of the pound-mass (1.0 slug = 32.174 lbm).

6.8 Conversion of Units

Four typical systems of mechanical units presently being used in the United States are listed in Table 6.10. The table does not provide a complete list of all possible quantities; this list is presented to demonstrate the different terminology that is associated with each unique system. If a physical quantity is expressed in any system, it is a simple matter to convert the units from that

Table 6.10 Mechanical Units

| Quality | Absolute System | | Gravitational System | |
	MKS	CGS	Type I	Type II
Length	m	cm	ft	ft
Mass	kg	g	slug	lbm
Time	s	s	s	s
Force	N	dyne	lbf	lbf
Velocity	$m \cdot s^{-1}$	$cm \cdot s^{-1}$	$ft \cdot s^{-1}$	$ft \cdot s^{-1}$
Acceleration	$m \cdot s^{-2}$	$cm \cdot s^{-2}$	$ft \cdot s^{-2}$	$ft \cdot s^{-2}$
Torque	$N \cdot m$	$dyne \cdot cm$	$lbf \cdot ft$	$lbf \cdot ft$
Moment of inertia	$kg \cdot m^2$	$g \cdot cm^2$	$slug \cdot ft^2$	$lbm \cdot ft^2$
Pressure	$N \cdot m^{-2}$	$dyne \cdot cm^{-2}$	$lbf \cdot ft^{-2}$	$lb \cdot ft^{-2}$
Energy	J	erg	$ft \cdot lbf$	$ft \cdot lbf$
Power	W	$erg \cdot s^{-1}$	$ft \cdot lbf \cdot s^{-1}$	$ft \cdot lbf \cdot s^{-1}$
Momentum	$kg \cdot m \cdot s^{-1}$	$g \cdot cm \cdot s^{-1}$	$slug \cdot ft \cdot s^{-1}$	$lbm \cdot ft \cdot s^{-1}$
Impulse	$N \cdot s$	$dyne \cdot s$	$lbf \cdot s$	$lbf \cdot s$

Type I—U.S. Customary
Type II—Engineering

system to the one in which you are working. To do this, you must know the basic conversion for the unit involved and must follow a logical series of steps.

Mistakes can be minimized if you remember that a conversion factor simply relates the same physical quantity in two different unit systems. For example, 1.0 in and 25.4 mm each describes the same length quantity. Thus when using the conversion factor 25.4 mm/in to convert a quantity in inches to millimeters, you are multiplying by a factor that is not numerically equal but is physically identical. This fact allows you to avoid the most common error, that of using the reciprocal of a conversion. Just imagine that the value in the numerator of the conversion must describe the same physical quantity as that in the denominator. When so doing, you will never use the incorrect factor 0.304 8 ft/m since 0.304 8 ft is clearly not the same length as 1 m.

Example prob. 6.2 demonstrates a systematic procedure to use when performing a unit conversion. The construction of a series of individual quantities will aid the thought process and help to ensure a correct unit analysis. In other words, the units to be eliminated will cancel out, leaving the desired results. The final answer should be checked to make sure it is reasonable. For example, the results of converting from inches to millimeters should be approximately 25 times larger than the original number.

Example problem 6.2 Convert 6.7 in to millimeters.

Solution Write the identity

$$6.7 \text{ in} = \frac{6.7 \text{ in}}{1}$$

then multiply this identity by the appropriate conversion factor.

$$6.7 \text{ in} = \frac{6.7 \text{ in}}{1} \frac{25.4 \text{ mm}}{1 \text{ in}} = 1.7 \times 10^2 \text{ mm}$$

Example problem 6.3 Convert 85.0 lbm/ft³ to kilograms per cubic meter.

Solution

$$85.0 \text{ lbm/ft}^3 = \frac{850 \text{ lbm}}{1 \text{ ft}^3} \frac{(1 \text{ ft})^3}{(6.304\ 8 \text{ m})^3} \frac{0.453\ 6 \text{ kg}}{1 \text{ lbm}}$$

$$= 1.36 \times 10^3 \text{ kg/m}^3$$

Example problem 6.4 Determine the gravitational force (in newtons) on an automobile with a mass of 3 645 lbm. The acceleration of gravity is known to be 32.2 ft/s².

Solution A Force, mass, and acceleration of gravity are related by
$F = mg_L/g_C$

$$m = \frac{3\ 645 \text{ lbm}}{1} \frac{1 \text{ kg}}{2.204\ 6 \text{ lbm}} = 1\ 653.36 \text{ kg}$$

$$g_L = \frac{32.2 \text{ ft}}{1 \text{ s}^2} \frac{0.304\ 8 \text{ m}}{1 \text{ ft}} = 9.814\ 6 \text{ m/s}^2$$

$$F = mg_L/g_C = (1\ 653.36 \text{ kg})(9.814\ 6 \text{ m/s}^2)/(1.0 \text{ kg m/N} \cdot \text{s}^2)$$

$$= 16\ 227 \text{ N} = 16.2 \text{ kN}$$

(*Note*: Intermediate values were not rounded to final precision, and we have used either exact or conversion factors with at least one more significant figure than that contained in the final answer.)

Solution B

$$F = \frac{mg_L}{g_C} = \frac{3\ 645 \text{ lbm}}{1} \frac{32.2 \text{ ft}}{1 \text{ s}^2} \frac{1 \text{ kg}}{2.204\ 6 \text{ lbm}} \frac{0.304\ 8 \text{ m}}{1 \text{ ft}} \frac{\text{N} \cdot \text{s}^2}{1.0 \text{ kg} \cdot \text{m}}$$

$$= 16\ 227 \text{ N} = 16.2 \text{ kN}$$

(*Note*: It is often convenient to include conversions with the appropriate engineering relationship in a single calculation.)

Example problem 6.5 Convert a mass flow rate of 195 kg/s (typical of the airflow through a turbofan engine) to slugs per minute.

Solution

$$195 \text{ kg/s} = \frac{195 \text{ kg}}{1 \text{ s}} \frac{1 \text{ slug}}{14.954 \text{ kg}} \frac{60 \text{ s}}{1 \text{ min}} = 782 \text{ slug/min}$$

Example problem 6.6 Compute the power output of a 225 hp (horsepower) engine in (*a*) British thermal units per minute and (*b*) kilowatts.

Solution

(*a*) $225 \text{ hp} = \dfrac{225 \text{ hp}}{1} \dfrac{2.546 \, 1 \times 10^3 \text{ Btu}}{1 \text{ hp} \cdot \text{h}} \dfrac{1 \text{ h}}{60 \text{ min}}$

$= 9.55 \times 10^3 \text{ Btu/min}$

(*b*) $225 \text{ hp} = \dfrac{225 \text{ hp}}{1} \dfrac{0.745 \, 70 \text{ kW}}{1 \text{ hp}} = 168 \text{ kW}$

The problem of unit conversion becomes more complex if an equation has a constant with hidden dimensions. It is necessary to work through the equation converting the constant K_1 to a new constant K_2 consistent with the equation units.

Consider the following example problem given with English units.

Example problem 6.7 The velocity of sound in air (*c*) can be expressed as a function of temperature (*T*):

$$c = 49.02 \sqrt{T}$$

where *c* is in feet per second and *T* is in degrees Rankine.

Find an equivalent relationship when *c* is in meters per second and *T* is in kelvins.

Procedure

1. First, the given equation must have consistent units; that is, it must have the same units on both sides. Squaring both sides, we see that

$$c^2 \text{ft}^2/\text{s}^2 = 49.02^2 T°\text{R}$$

It is obvious that the constant $(49.02)^2$ must have units in order to maintain unit consistency. (The constant must have the same units as c^2/T.)

Solving for the constant, we get

$$(49.02)^2 = \frac{c^2 \text{ ft}^2}{s^2} = \left[\frac{1}{T°\text{R}}\right] = \frac{c^2}{T}\left[\frac{\text{ft}^2}{s^2°\text{R}}\right]$$

2. The next step is to convert the constant $49.02^2 \text{ ft}^2/(\text{s}^2\text{R})$ to a new constant that will allow us to calculate *c* in meters per second given *T* in kelvins.

We recognize that the new constant must have units of square meters per second squared-kelvin.

$$\frac{(49.02)^2 \text{ ft}^2}{1 \text{ s}^2 {}^\circ R} = \frac{(49.02)^2 \text{ ft}^2}{1 \text{ s}^2 {}^\circ R} \frac{(0.3048 \text{ m})^2}{(1 \text{ ft})^2} \frac{9{}^\circ R}{5 \text{ K}} = \frac{401.84 \text{ m}^2}{1 \text{ s}^2 \text{K}}$$

3. Substitute this new constant 401.84 into the original equation

$$c^2 = 401.84 \ T$$

$$c = 20.05 \ \sqrt{T}$$

where c is in meters per second and T is in kelvins.

Key Terms and Concepts

The following are some of the terms and concepts you should recognize and understand.

Physical quantities	**Fundamental dimensions**
Units	**Derived dimensions**
SI units	**Absolute system**
Base units	**Gravitational system**
Supplementary units	**Metric system**
Derived units	**Symbols**
Dimensions	**U.S. Customary System**
Dimensional system	**Engineering System**

Problems

6.1 Using the correct number of significant figures, convert the following physical quantities into the proper SI units.
- (a) 725 lbm
- (b) 98.6°F
- (c) 3.20×10^2 acres
- (d) 52×10^3 gal
- (e) 231.0×10^3 gal/h
- (f) 62 ft/s
- (g) 235 hp
- (h) 2020 in
- (i) 1.255×10^3 ft^3/min
- (j) 25 slugs

6.2 Convert the following to SI units. Use the correct significant figures.
- (a) 650.5 Btu/min
- (b) 55.2 hp · h
- (c) 4.250×10^3 mi
- (d) 2.00×10^2 mi/h
- (e) 235 lbf
- (f) 8.450×10^4 lbm/ft^3
- (g) 1.920 atm
- (h) 242°F
- (i) 5280.0 ft
- (j) 64 oz

6.3 Convert as indicated giving the answer with proper significant figures.
- (a) 72.5 in to centimeters
- (b) 525 L to cubic feet
- (c) 65.9 C to degrees Fahrenheit
- (d) 10.5×10^3 bushels to cubic centimeters
- (e) 50.7×10^5 Btu/h to kilowatts

6.4 Convert as indicated giving the answer with proper significant figures.
- (a) 7.450×10^3 km to feet
- (b) 295 K to degrees Fahrenheit
- (c) 5.75×10^4 ft · lbf to joules
- (d) 14.7 lbf/in^2 to pascals
- (e) 62.6 slug/ft^3 to grams per cubic centimeters

6.5 Using the rules for expressing SI units, express each of the following in correct form if given incorrectly:

(a) 475 n

(b) 8 000 pa

(c) 62.5 j

(d) 11.5 cm's

(e) 5000 K

(f) 100.0 m m

(g) 9.5 m/s/s

(h) 25 amp

(i) 300 degrees Kelvin

(j) 115 kW/hours

6.6 Using the rules for expressing SI units, express each of the following in correct form if given incorrectly:

(a) 101 C

(b) 1.000 N

(c) 40 nM

(d) 53 m per sec

(e) 25 farads

(f) .5 m · m

(g) 75 A's

(h) 48 Kg

(i) 8050 N/m/m

(j) 5226 J/sec

6.7 If a force of 1.25×10^3 N is required to lift an object with a uniform velocity and an acceleration of gravity shown as follows, determine the mass, in kg, of the object:

(a) 32.5 ft/s^2

(b) 9.86 m/s^2

(c) 22 ft/s^2

6.8 Determine the acceleration of gravity required (meters per second squared) to lift a 1300 kg object at uniform velocity when the force required to lift it is

(a) 2.850×10^3 lbf

(b) 2395 lbf

(c) 9.90×10^3 N

6.9 What work is done to lift a 9.50×10^2 kg object 3.25×10^1 ft vertically when the acceleration of gravity is 32.06 ft/s^2? Express your answer in joules. (*Note:* Work = force × distance traveled in the force direction.)

6.10 A train is traveling at 5.00×10^1 meters per second. The resistance between the train and the track is 73.0 newtons. If air resistance can be ignored, determine the horsepower needed to power the engine. [*Note:* Power = (force)(velocity)]

6.11 The average density of Styrofoam is 1.00 kg/m^3. If a Styrofoam cooler is made with outside dimensions of $3.00 \times 2.00 \times 1.50$ ft and inside dimensions of $2.50 \times 1.50 \times 1.00$ ft, what is the volume of the Styrofoam used in cubic meters? What is the mass in kilograms?

6.12 A hemispherical dome is filled with hydrogen (density = $8.375 \times 10^{-5} \text{ g/cm}^3$) at 20 C and has an inside base diameter of 1.000×10^4 ft. What is the volume inside the dome, in m^3? What is the mass of the hydrogen, in kg?

6.13 A backyard above-ground pool measures 5.0 m in diameter. The water depth is 2.0 m. The atmospheric pressure on the top of the water is 14.7 psi (pound per square inch). Determine the total force exerted by the water on the ground if the pressure increases as $P = 14.7 + h(\rho)$. The density of the water is 62.4 lbm/ft^3. Express your answer in lbf.

6.14 Assume your hometown is growing so rapidly that another water tower is necessary to meet the needs of the community. Engineers predict that the water tower will need to hold 1.00×10^6 kilograms of water, with a density of 999 kilograms per cubic meter.

(a) If this tower is 50.0 meters high and spherical, what will the volume of the tower have to be?

(b) Determine the vertical force, in kN, acting on each of the three evenly spaced legs from the weight of the water alone.

(c) What would be the diameter of the tower?

6.15 In the early 1900s the city of Chicago had a 150 year blizzard. To remove the snow from the streets, the snow was piled on frozen Lake Michigan. Assuming that the snow stood on the frozen lake in a perfect cone with a base diameter of 200.0 ft and a height of 18.00 ft, determine

(a) the volume of the pile

(b) the mass of the snow (density is 850 kg/m³)

(c) the force that the snow exerted on Lake Michigan, in N.

6.16 A pile of sand in a perfect cone has a height of 20.0 ft and a diameter of 100.0 ft. If this sand has a density of 97.0 lbm/ft³,

(a) determine the volume of the cone

(b) determine the mass of the sand in the cone

(c) during winter months the sand will be spread on county roads. Using a spreadsheet, determine the volume and mass of the sand remaining if the height decreases from 20.0 to 5.00 ft in increments of 0.500 ft. Assume that the base remains constant.

6.17 Three tree trunks, 20 ft long, each with a constant diameter of 14 in, are cut down. Find

(a) their total volume

(b) the mass of each tree trunk if the density of each is 600.0 kg/m³ (kilograms per cubic meter).

(c) the weight of the three tree trunks combined, in N.

6.18 If you were on another planet, say, Mars, which of the following, g_C or g_L, would change and which would stay constant? Explain the difference.

6.19 The ideal gas law shows the relationship among some common properties of ideal gases.

$$pV = n\mathbf{R}T$$

where p = pressure

V = volume

n = number of moles of the ideal gas

$\mathbf{R}$ = universal gas constant = 8.314 kJ/(kmol · K)

T = absolute temperature

If you have 3 moles of an ideal gas at twenty degrees Celsius and it is stored in a container that is 0.500 meters on each side, calculate the pressure in Pa.

6.20 A small portable pressure tank can be pressurized using an air compressor. The tank is a cylinder with two hemispherical ends. Its inside dimensions are 12 inches in diameter and 36 inches end to end. The maximum tested pressure that can be maintained safely at 70°F is 200 psi (pounds per square inch) and it will fly apart at 250 psi.

(a) Determine the inside volume of the pressure tank.

(b) Calculate the specific volume (v = V/m) at 70°F and 200 psi.

(c) Assume that the tank has been filled using the mass of air in part (b). What is the maximum temperature the tank can withstand?

(d) With the aid of a spreadsheet, show the pressure range the tank in part (b) would experience if the temperature varies from −40 to 190°F in 5 degree increments. Hint:

$Pv = RT$

Pressure, P—lbf/ft²

Sp. vol, v—ft³/lbm

Gas constant (R_{air} = 53.33 ft · lbf/lbm°R)

Temperature, T—Must be absolute (°R)

6.21 The change in pressure across a partial blockage in an artery, called a stenosis, can be approximated by the following equation, where

$$p_2 - p_1 = K_v\,(\mu)\,\frac{V}{D} + K_u\left(\frac{A_0}{A_1} + 1\right)^2 \rho(V^2)$$

where V = blood velocity

μ = blood viscosity, $N \cdot s/m^2$

density = blood density, in kg/m^3

D = artery diameter

A_0 = area of unobstructed artery

A_1 = area of the stenosis

Determine the dimensions of the constants K_v and K_u in both the SI System and the Engineering System.

6.22 A force, **P**, exerted on a spherical particle can sometimes be described as $\mathbf{P} = 3\pi(\mu DV)$.

where μ = the fluid viscosity $(N \cdot s/m^2)$

D = particle diameter

V = particle velocity

What are the dimensions of the constant 3π? Will the constant change with a different unit system?

6.23 A rectangular raft with dimensions of 3 ft in width, 6 ft in length, and 8 in depth is fully inflated. The raft is made of a 0.125 inch thick rubber with a density of 640 kilograms per cubic meter. Assume that the weight of air inside is negligible. Determine
(a) the outside surface area, in ft^2
(b) the mass of the raft, in kg
(c) if the raft will float in water; (water density = 62.4 lbm/ft^3 pound-mass per cubic foot)
(d) how much weight could be placed on the raft before it would sink?

6.24 A cylindrical underground storage tank with a radius of 10.0 ft and a height of 40.0 ft is filled with gasoline. Given a density of 670 kg/m^3, determine the mass of the gasoline in the tank. If the average automobile gas tank holds 25.0 gal, compute the number of autos that may be filled from this underground facility.

6.25 A cylindrical tank 20.0 ft long and 10.0 ft in diameter is oriented so that its longitudinal axis is horizontal. Develop a table that will
(a) Show how many gallons of diesel fuel are in the tank if the fluid level is measured in 1.00 ft increments from the bottom of the tank
(b) Show the corresponding mass at each increment, in kg, if its specific gravity is 0.73.

6.26 An object in the shape of a right cone with a base diameter of 275 mm and a height of 325 mm is floated point-up in water (density of 62.4 lbm/ft^3). The point of the cone extends 115 mm above the surface of the water (see Fig. 6.4).
(a) What is the density of the cone, in kg/m^3?
(b) What is the mass of the cone, in kg?

6.27 Construction workers accidentally dropped a small spherical piece of a new composite material with a diameter of 3 inches into a bucket of engine oil. If the engine oil has a density of 55.2 lbm/ft^3 (pound-mass per cubic foot) and the composite material has a density of 34 lb/ft^3, determine
(a) the outside surface area of the spherical piece of composite material, in square inches

Figure 6.4

225

Problems

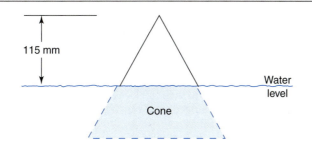

115 mm

Water
level

Cone

(b) the mass, in pound-mass
(c) whether the sphere will float
(d) how far it will sink into the oil if it does float

6.28 A weir is used to measure flow rates in open channels. For a rectangular weir the expression can be written $Q = 288.8\ LH^{3/2}$ (see Fig. 6.5).

where Q = discharge rate, in gal/h

L = length of weir opening parallel to liquid, in inches

H = height of fluid above crest, in inches

(a) As the channels become larger, the weir opening can be expressed as follows: Q in cubic feet per second, with L and H in feet. Determine the new constant.

(b) Write a computer program or prepare a spreadsheet to produce a table of values of Q in both ft^3/s and gallons per hour, with L (inches) and H (inches). Use values of L that range from 1 to 36 inches in 1 inch increments. Let H range from 2 to 48 in increments of 2 inches.

6.29 Conservation of energy suggests that potential energy is converted to kinetic energy when an object falls in a vacuum. Velocity at impact can be determined as follows:

$$V = 5.47\ \sqrt{h}$$

where V is velocity, in mph, and h is distance, in ft.

(a) Determine a new constant so that the equation is valid for h in ft and V, in feet per second (ft/s).

Figure 6.5

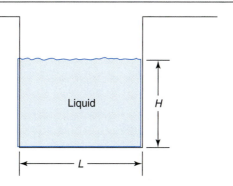

Liquid

H

L

(b) If you drop an object from 145 ft, what is its velocity on impact with the ground, in mph and ft/s?

6.30 Las Vegas developers want to construct an Egyptian-style right pyramid, with a square base measuring 210 ft on each side and a height of 210 ft, to attract gamblers to a new casino. They plan to construct the pyramid out of structural steel with a brick facade (brick density of 1800 kg/m^3). The desired bricks are available in 2, 4, 6, 8, 10, 12, 14, 16, and 18 × 18 × 10 inches; that is, 2 × 18 × 10, 4 × 18 × 10, and so on.

(a) Using as many 18 × 18 × 10 inch bricks as possible, determine the exact number of each brick size needed to construct the pyramid. (*Hint*: Develop a spreadsheet or table as follows.)

Elevation Above Base, inches	# 18 × 18 × 10 inch Bricks	Others	Qtn
0	556	None	
10	552	8 × 18 × 10	4
20	548	16 × 18 × 10	4
And so on			

(b) Compute the mass of the bricks required to construct the pyramid, in tons.

Answers to Selected Problems

Chapter 3

3.2 Angle $B = 20.3°$; angle $C = 35.7°$; side $a = 12.5$ cm
3.6 (a) 20 knots; (b) $V_1 = 299$ ft/s; $\alpha = 19.2°$
3.12 Distance = 1.22 mi., one trip
3.17 48.57 s
3.22 (a) **13 790** rev; (b) 5 861 rev; (c) mountain; 457%, touring: 475%
3.25 (a) 0 m/s; (b) 11.5 m; (c) 15 m/s downward; (d) 31.8 m/s; (e) 4.77 s

Chapter 4

4.2 Plotting exercise
4.4 Plotting exercise
4.6 (b) Velocity = $1.6t - 1.8$
   (c) Acceleration = 1.6 ft/s^2
4.8 (b) $T = 19(\text{emf}) - 14$
4.10 Plotting exercise
4.12 I = 0.02 V + 0.06
4.14 Parts = 20.1 $e^{0.25\,(\text{month})}$
4.16 $V = 3.1416\,R^2$
4.18 $V = 4.4217h^{0.4977}$
4.20 (a) Downloads = $15.344e^{0.1985(\text{day})}$
   (b) Day 30 = 5920
4.22 (a) 4991.25 Btu/ft^3
   (b) 51.840 Btu/ft^3
   (c) 550.32°F
   (d) 1100.64°F

Chapter 5

5.2 (c) 2; (f) exact conversion; (h) 4
5.5 (a) 44; (c) $128 000
5.6 (b) 61–69 lb/in^2

Chapter 6

6.2 (a) 1.144×10^4 W
   (b) 14.8×10^7 kg·m^2/s^2
   (c) 6.840×10^6 m
   (d) 89.4 m/s
   (e) 1.05×10^3 kg·m/s^2
   (f) 1.354×10^6 kg/m^3
   (g) 1.95×10^5 kg/m·s^2
   (h) 390 K
   (i) 1609.3 m
   (j) 1.8 kg
6.4 (a) 2.444×10^7 ft
   (b) 71.3°F
   (c) 7.78×10^2 N·m
   (d) 1.01×10^5 Pa
   (e) 32.3 g/cm^3

6.6 (a) 101° C

(b) 1000 N

(c) 40 N · m

(d) 53 m/s

(e) 25 F

(f) 0.5 mm

(g) 75 A

(h) 48 kg

(i) 8050 N/m^2

(j) 5226 J/s

6.8 (a) $g_L = 9.752$ m/s^2

(b) $g_L = 8.195$ m/s^2

(c) $g_L = 7.62$ m/s^2

6.10 Power = 4.89 hp

6.12 (a) Volume = 7.413×10^6 m^3

(b) Mass = 6.209×10^5 kg

6.14 (a) Volume = 1001 m^3

(b) Force = 3.27×10^6 N

(c) Diameter = 12.4 m

6.16 (a) Volume = 52 400 ft^3

(b) Mass = 5.08×10^6 lbm

(c) Spreadsheet

6.18 (a) g_L = never changes

(b) g_L = depends on distance from center of mass

6.20 (a) Volume = 2.1 ft^3

(b) Specific vol. = 0.98 ft^3/lbm

(c) $t = 202.5°$F

(d) Spreadsheet

6.22 (a) 3 pi is dimensionless

(b) It will not change.

6.24 (a) Mass = 526 000 lbm

(b) No. of Autos = 3 760

6.26 (a) 956 kg/m^3

(b) 6.15 kg

6.28 (a) Constant = 5.35

(b) Spreadsheet

6.30 (a) Spreadsheet

(b) Mass = 7357 tons

Index

Abscissa, 152
Accreditation, 63
Accuracy, 187
Alternative solutions, 98–102
Analysis, 5, 102–107
Approximations, 190–198
Axes breaks, 152
Axis labeling, 158

Bill of materials, 115
Bloom's taxonomy, 85
Brainstorming, 99

Cause-and-effect diagrams, 45
Checkoff lists, 98
Code of ethics, 69
Communication, 115–120
Constraints, 94–95
Craftsperson, 10–11
Criteria, 95–97
Crosby, Philip, 40
    Crosby's 14 steps, 40
Curve fitting, 163
Customer, 43

Data recording, 145, 149
Decision, 108–113
Deming, W. Edwards, 38
    Deming's 14 points, 39
Design, 5, 79
Design process, 5, 80
Dimensional systems, 203
    absolute, 203
    gravitational, 203
Dimensions, 202

Empirical equations, 164
    exponential, 172
    linear, 164
    power, 166
Empirical functions, 163
Engineer
    definition, 4
    description of, 4–6
    education of, 61–66
Engineering design
    activity time schedule, 82

Engineering design—*Cont.*
    conceptual, 82
    customer satisfaction, 84–86
    definition of, 80
    optimization, 83, 108
    oral presentation, 118–120
    preliminary, 83
    process, 80
    prototype, 83
    written report, 116–118
Engineering disciplines, 23–24
    aerospace, 24–26
    chemical, 26
    civil, 27–29
    computer, 29–30
    electrical, 29–30
    environmental, 30–31
    industrial, 31–33
    mechanical, 33–34
Engineering functions, 11–23
    construction, 18
    consulting, 22
    design, 15–16
    development, 13–15
    management, 21
    operations, 18–20
    production, 17–18
    research, 12–13
    sales, 20–21
    teaching, 22–23
    testing, 17–18
Engineering method, 126
Engineer's Creed, 69

Facilitator, 52
Functional scales, 155

Graph paper, 152
Graphical analysis, 148
Graphical presentation, 147
Graphing procedures, 151

Identification of a need, 86–88
International System of Units (SI), 205
Ishikawa, Kaoru, 43

Junkins, Jerry, 42

Method of least squares, 164,
Method of selected points, 164
Myers-Briggs type indicator (MBTI), 58

Ordinate, 152

Pareto chart, 44
Payoff function, 109
Personal style inventory, 57
Physical quantities, 201
Plotting symbols, 159
Precision, 187
Problem analysis, 125–126
    art, 126
    science, 125
Problem definition, 88–90
Problem layout, 129–137
Problem presentation, 128
Problem solving models, 47
Problem solving process, 125
Problem space, 49
Professional registration, 68–69

Random errors, 189
Recorder, 52
Reverse engineering, 91

Scale calibrations, 155
Scale graduations, 155
Scientific notation, 184
Scientific problem-solving method, 3, 128
Scientist, 3
Search, 90–94
Significant digits, 183–187

Solution documentation, 130
Solution space, 49
Specification, 113–115
Sponsor, 51
Synthesis, 6, 86
System International d'Unites (SI), 205
Systematic errors, 188–189

Taylor, Frederick, 37
Team, 51
Team dynamics, 52–54
    forming, 53
    norming, 53–54
    performing, 54
    storming, 53
Team leader, 51–52
Team meetings, 54–57
Technical spectrum, 3
Technician, 10
Technologist, 10
Technology team, 2–3
Total quality, 34
    definition of, 41
    elements of, 42
Trade-offs, 108

Unit conversions, 217–221
Unit prefixes, 209
Units, 204
    Engineering System, 216
    metric (SI), 205–215
    metric system rules, 210–215
    U. S. Customary System, 215